CW00381658

PRACTICAL SUSTAINABILITY STRATEGIES

PRACTICAL SUSTAINABILITY STRATEGIES

HOW TO GAIN A COMPETITIVE ADVANTAGE

Nikos Avlonas
Founder and President
Center for Sustainability & Excellence

George P. Nassos
Principal
George P. Nassos & Associates, Inc.

Copyright © 2014 by John Wiley & Sons, Inc. All rights reserved

Published by John Wiley & Sons, Inc., Hoboken, New Jersey

Published simultaneously in Canada

No part of this publication may be reproduced, stored in a retrieval system, or transmitted in any form
or by any means, electronic, mechanical, photocopying, recording, scanning, or otherwise, except as
permitted under Section 107 or 108 of the 1976 United States Copyright Act, without either the prior
written permission of the Publisher, or authorization through payment of the appropriate per-copy fee to
the Copyright Clearance Center, Inc., 222 Rosewood Drive, Danvers, MA 01923, (978) 750-8400, fax
(978) 750-4470, or on the web at www.copyright.com. Requests to the Publisher for permission should
be addressed to the Permissions Department, John Wiley & Sons, Inc., 111 River Street, Hoboken, NJ
07030, (201) 748-6011, fax (201) 748-6008, or online at http://www.wiley.com/go/permission.

Limit of Liability/Disclaimer of Warranty: While the publisher and author have used their best
efforts in preparing this book, they make no representations or warranties with respect to the accuracy
or completeness of the contents of this book and specifically disclaim any implied warranties of
merchantability or fitness for a particular purpose. No warranty may be created or extended by sales
representatives or written sales materials. The advice and strategies contained herein may not be suitable
for your situation. You should consult with a professional where appropriate. Neither the publisher nor
author shall be liable for any loss of profit or any other commercial damages, including but not limited
to special, incidental, consequential, or other damages.

For general information on our other products and services or for technical support, please contact our
Customer Care Department within the United States at (800) 762-2974, outside the United States at
(317) 572-3993 or fax (317) 572-4002.

Wiley also publishes its books in a variety of electronic formats. Some content that appears in print may
not be available in electronic formats. For more information about Wiley products, visit our web site at
www.wiley.com.

Library of Congress Cataloging-in-Publication Data

Avlonas, Nikos, 1969–
Practical sustainability Strategies: How to Gain a Competitive Advantage / Nikos Avlonas,
Founder and President Center for Sustainability & Excellence, George P. Nassos,
Principal George P. Nassos & Associates, Inc.
    pages cm
  Includes index.
  ISBN 978-1-118-25044-0 (cloth)
  1. Business enterprises–Environmental aspects.   2. Industrial management–Environmental
aspects.   3. Sustainable development.   4. Social responsibility of business.
I. Nassos, George P.   II. Title.
  HD30.255.A96 2014
  658.4′083–dc23
                                                                2013021899

Printed in the United States of America

10 9 8 7 6 5 4 3 2 1

CONTENTS

About 35 years ago, I had the fortune to work for a subsidiary of an American chemical company that was located in Cologne, Germany. While experiencing many great times during the three years I lived in Europe, there were a few that stood out because of their impact on the natural environment and energy efficiency. Here are three that I will never forget.

1. I was walking to the subway and had to take the escalator to the lower level. When I arrived for the first time, I was disappointed that the escalator was not operating. So I did what I would normally do under such a circumstance—walk down the stationary escalator. As soon as I took the first step to walk down, the escalator started and continued to operate until I reached my destination. With no one else on the escalator, it stopped when I got off. There was a pressure switch under the first step that activated the motor. That really made sense.

2. I was participating in a business meeting at a high-rise office building in the La Defense suburb of Paris. Sitting in a conference room on about the 40th floor with nothing but windows on the outside wall, I could not help but notice that from time to time the lights would go off and later come back on. After witnessing this occurrence two or three times, I asked the hosts if they were having problems with their electrical supply. I was embarrassed to learn that light sensors controlled the need for artificial light when the sunlight was insufficient. This activity on that day was due to a partly cloudy day.

3. I lived in an apartment building two blocks from the Rhine River and about 6 or 7 kilometers south of the city center where my office was located. My commute to the office was by driving on a four-lane boulevard along the river. About every kilometer or less were intersections with stoplights to allow traffic to enter or exit the boulevard. In addition to these traffic lights at the intersections, digital signals existed on the boulevard about midway between the intersections. These digital signals would indicate what speed the cars should travel in order to arrive at the next intersection with a green light. This system provided several benefits like reduced fuel consumption, reduced emissions, reduced

congestion, and reduced stress. I am not sure how many years this system was in operation prior to my experiencing it. But it surely made sense.

After working for the chemical company for 16 years, I returned to the United States and joined an environmental company for the next 15 years. A combination of my experiences in Europe, and Germany in particular, along with working for an environmental company convinced me that this was a field of great interest. I then decided to enter the academic field and combine my chemical and environmental background with my business background (I also have an MBA) and teach at a graduate business school. I wanted to offer students a combination of teaching theory as well as my business experiences in the real world. I was fortunate to learn that the Illinois Institute of Technology–Stuart School of Business had just developed a new program, MS in Environmental Management, and the dean was looking for a director of the program. This was a great fit for both the school and me.

During my first year at the school in 1997, I attended a conference sponsored by the World Resources Institute focusing on environmental and social sustainability. About 150 people attended this annual conference with about 70% from academia and the balance from the business and government sectors. I had never heard of this topic, but I found it extremely interesting. The following year I attended the conference again, and after this second experience I was convinced that sustainability was very, very important. During the next academic year, I introduced sustainability into the capstone course I was teaching and renamed it "Business Strategy: The Sustainable Enterprise." Over the next few years, I introduced the sustainability concept in other courses and shortly thereafter changed the program name to MS in Environmental Management and Sustainability. This program was eventually ranked No. 11 in the world by the Beyond Grey Pinstripes biennial survey.

I became so interested in sustainable strategies that I researched the topic on a continuous basis and added new strategies and case studies to the course almost every year. Since about 80% of the students were working professionals and part-time students, the final exam for the course was actually a project. Each student was challenged to apply one of the sustainable strategies to their workplace and show how it would enhance the company, a subsidiary or a strategic business unit in terms of environmental integrity, social equity, and economic viability. If it did not, I did not want to hear about it. The full-time students had the option to select a publicly held company or create a new one based on the sustainable strategy.

In addition to researching for new strategies, I was also interested in books about sustainability. Over the past 10 years, I did not come across any book that dealt with the various sustainable strategies I was teaching. I am convinced that sustainability should not be a discipline in a business school like marketing, accounting, finance, or organizational behavior, but rather should be imbedded in all the appropriate courses. All graduates of a business school should have an understanding of the sustainability concept and its benefits. This book was written to enhance business programs at any or all business schools. It can also be used as the basis for a course on sustainability or as a reference to cover the topic in one or two modules of any other business course.

This does not mean that the book has been written only for business schools. Small- and medium-sized companies as well as large corporations will certainly benefit from the contents of this book. In addition, government agencies and nongovernmental organizations will also benefit by adopting selective strategies. Just as I am convinced that sustainability should not be a discipline in a business school, it should not be the responsibility of an individual or department in a company or organization. Sustainability should be imbedded in the culture of the organization so everyone can work together to achieve their goal of operating as a truly sustainable company.

While reading many books on sustainability, one that can be considered a classic is *The Limits to Growth* written in 1972 by Donella H. Meadows, Dennis L. Meadows, and Jergen Randers [1]. They used a computer model to simulate the consequence of interactions between humans and the Earth's ecological systems to predict the state of our environment through the year 2100. In 2004, the authors wrote *Limits to Growth: The 30-Year Update* [2] as a review of the computer model. They showed that the actual data relate very closely to the model's prediction, and if the real data continue to follow the model, the world may be heading to a collapse. This was followed with a 40-year anniversary meeting of experts leading to another paper published in *Smithsonian Magazine* [3]. Graphic design by Linda Eckstein showing what has happened during these 30+ years has been reproduced on the cover of this book because it reflects another inspiration to write this book.

A few years ago, I had the pleasure of meeting Nikos Avlonas, CEO and Founder of the Center for Sustainability and Excellence, an organization that trains and consults in the sustainability field. His expertise is in the measurement of sustainability metrics that are used to determine how well the organization is performing and in the reporting of its activities. Consequently, it made sense to join forces and write a book about the strategies, measurements, reporting, and communicating.

GEORGE P. NASSOS
2013

REFERENCES

1. Meadows DH, Meadows DL, Randers J. *The Limits to Growth*. New York: Universe Books; 1972.
2. Meadows DH, Meadows DL, Randers J. *Limits to Growth: The 30-Year Update*. White River Junction, VT: Chelsea Green Publishing Company; 2004.
3. Strauss M. "Got Corn?" *Smithsonian Magazine*; April 2012.

To the esteemed Dr. George P. Nassos, President of Sustainable Energy Systems, our beloved son in the Lord: grace and peace on high.

It is with sincere joy that we write to congratulate you on the publication of your book *Practical Sustainability Strategies: How to Gain a Competitive Advantage*, coauthored with your colleague Nikos Avlonas, President of the Center for Sustainability and Excellence.

As you are aware, for over twenty years, we have organized international symposia, summits and seminars, drawing attention to the need for approaching the need for ecological crisis on an inter-disciplinary basis, involving leaders of the scientific, religious, political, and corporate worlds. We recall your participation as one of our invited guests at the successful Halki Summit in June 2012.

Therefore, we are delighted that you are sharing these values in the business and economic domains, promoting the critical urgency of environmental sustainability and offering practical and sustainable strategies and systems, assessment and engagement, as well as resources and communication.

It is our sincere hope and prayer that you may continue to advocate this respectful approach and management by human beings as stewards of God's creation.

At the Ecumenical Patriarchate, the 3rd September, 2013

Your fervent supplicant before God,

†**BARTHOLOMEW**
Archbishop of Constantinople-New Rome
and Ecumenical Patriarch*

*Ecumenical Patriarch of over 250 million Orthodox Christians worldwide. Given the name "Green Patriarch" by U.S. Vice-President Al Gore in 1997. http://www.firstpost.com/topic/person/bartholomew-i-vp-al-gore-the-green-patriarch-video-yTA0fmEHit0-47029-10.html

PART 1

INTRODUCTION TO SUSTAINABILITY

CHAPTER 1

Urgency to Adopt Sustainability

It has been about 50 years ago when we started to read books or articles about the environment with *Silent Spring* by Rachel Carson being one of the first important books published in 1962. Many other outstanding books have been written about the environment since then such as *The Ecology of Commerce* by Paul Hawken in 1993 and *Natural Capitalism* by Paul Hawken, Amory Lovins, and L. Hunter Lovins in 2008. The number and frequency of new books has increased as more and more people are concerned about the state of the environment.

Very few people question the decline in the state of our environment, only the degree to which it has deteriorated or the rate at which it is continuing to deteriorate. Regardless of the current status of our environment, it is important to put in perspective what has happened to our Earth since its creation. Historians estimate that the Earth is about 4.5 billion years old, but it is really difficult to understand exactly what this means. What does one billion really mean? Let's consider a situation where a 21-year-old girl is given one billion dollars as a gift, and she places the money in a noninterest-bearing account. She will be able to spend $60,000 every day of her life until she retires at the age of, say, 65 and still have $36 million left over for retirement. This gives someone a better understanding of what one billion dollars really means.

So how can we put 4.5 billion years in perspective so we can understand what has happened to the Earth since its creation. As suggested by David Brower [1], former executive director of the Sierra Club, let us compress the geological time, from the initial formation of the Earth until now, into the six days of biblical creation [2], from Monday through Saturday.

Practical Sustainability Strategies: How to Gain a Competitive Advantage,
First Edition. Nikos Avlonas and George P. Nassos.
© 2014 John Wiley & Sons, Inc. Published 2014 by John Wiley & Sons, Inc.

CREATION OF THE ENVIRONMENT

Using the compressed time scale, the Earth was formed at midnight, the beginning of the first day, Monday. There is no life until Tuesday about 8:00 A.M., and millions of species begin to appear and disappear throughout the week. Photosynthesis begins and it gets into high gear by Thursday morning, just after midnight. By Saturday, the sixth and last day of creation, there is sufficient oxygen in the atmosphere that amphibians can come onto land and enough chlorophyll manufactured for the vegetation to begin to form coal deposits. The giant reptiles appear around 4:00 P.M. and primates show up at 10:00 P.M. on this last day, but *Homo sapiens* don't appear until 11:59:54—just six seconds ago. In other words, if we compress the age of the Earth to six days, or 144 hours, "man" is not created until the last six seconds. A quarter of a second to midnight, Jesus Christ appears. One-fortieth of a second is the beginning of the industrial age, and one-eightieth of a second ago, we discover oil, thus accelerating the carbon blowout started by the industrial revolution.

Scientists have predicted that this 4.5 billion-year-old Earth will be around for another "week." But look at the damage that has been done in just the past one-fortieth of a second. About 70% of the major fisheries have been depleted or are at their biological limit. It is estimated that the forest cover has been reduced by as much as 50% worldwide; 50% of the wetlands and more than 90% of the grasslands have been lost [3]. Currently, almost 40% of the world's population is experiencing serious water shortages. The big question now is how long will we last, another one-fortieth of a second—about five generations? Or will we be able to survive for another quarter of a second—about 2000 years? Or can we make a difference to extend a healthy world to some indefinite period of time? Or is it too late, and are we in the midst of a period of overshooting the carrying capacity of the Earth, followed by a rapid collapse?

God did not create the natural environment for the benefit of the people so they can use and misuse it. The environment *can* be used indefinitely as long as it is replenished. It has the capacity to support the needs of living creatures—plants and animals, including humans—but only a finite number. If this carrying capacity is exceeded to the degree that it cannot be replenished, the population that it is supporting will decrease significantly. This can be demonstrated by a real experiment conducted by scientists a number of years ago.

EXCEEDING THE ECOLOGICAL FOOTPRINT

Near the end of World War II in 1944, the US Coast Guard placed 29 reindeer on St. Matthew Island in the Bering Sea as an emergency food supply for the US military. This island consisted primarily of vegetation and was void of any predators. Specialists had calculated that the island could support between 13 and 18 reindeer per square mile, or a total population of between 1600 and 2300 animals.

By 1957, the population was 1350; but just 6 years later in 1963, the population had exploded to 6000. Were the scientists wrong in their calculations of how many reindeer the island could support?

Eventually, it was determined that the original calculations had been correct. The 6000 reindeer vastly exceeded the carrying capacity of the island, and they were soon decimated by disease, starvation, and extreme weather conditions. Such a drastic overshoot, however, did not lead to restabilization at a lower level, with some of the reindeer dying off. Instead, the entire habitat was so damaged by the overshoot of reindeer that the number of animals fell far short of the original carrying capacity. By 1966, just three years later, there were only 42 reindeer living on St. Matthew Island rather than the expected 1600 to 2300.

This is an example of what could happen to the Earth. In the case of St. Matthew Island, the resources used by the reindeer were grasses, trees, and shrubs, all renewable resources that can be replenished. Many of the resources necessary for human survival, however, are not renewable. There is only a finite source of resources such as minerals, oil, and coal. We must be cognizant of the overutilization of both renewable and nonrenewable resources.

To examine this overutilization of the Earth's resources, we must look at a concept called the ecological footprint. This is a tool for measuring and analyzing human natural resource consumption and waste output within the context of nature's renewable and regenerative capacity (or biocapacity). It represents a quantitative assessment of the biologically productive area required to produce the resources (food, energy, and materials) and to absorb the wastes of an individual or region. In terms of resources, it includes cropland, grazing land, forest, fishing grounds, and built-up land. The footprint to handle waste output includes the forests required to absorb all the carbon dioxide emissions resulting from the individual's energy consumption.

In order to be sure we don't exceed the carrying capacity of the Earth, the footprint for humanity must be within the annual regenerative capability of nature. Similarly, we must not exceed the absorptive capacity of the planet for handling of the waste that is produced. A *sustainable* environment will exist if we live within the Earth's regenerative and absorptive capacity. If we remove more from nature than can be provided indefinitely, we are on an unsustainable track.

An organization called Global Footprint Network [4] has been calculating and analyzing the ecological footprint of about 140 countries. The footprint refers to the amount of the Earth's carrying capacity it takes to sustain humanity's consumption of goods and services, basically the need for food, clothing, shelter, energy, and disposal of waste. According to its calculations, in the late 1970s, humanity's collective ecological footprint breached the sustainability

TABLE 1.1   Ecological footprint for 2007

Country	Ecological footprint (acres)
United Arab Emirates	26.4
United States	19.8
Germany	12.6
Japan	11.6
Mexico	7.4
World	6.7
China	5.4
Biocapacity	4.4
India	2.2
Puerto Rico	0.1

Source: Global Footprint Network.

mark for the first time, and it has remained unsustainable ever since. In fact, the deficit for maintaining sustainability has grown every year since then, and it appears that this deficit is on a path to grow further in the foreseeable future. Currently, it is estimated that we need 1.4 Earths to insure that future generations are as well off as we are today [5].

It is interesting to note the variation in the ecological footprint by region or nation as seen in Table 1.1 [4]. Not surprising, the largest footprint belongs to the United Arab Emirates where it is 26.4 acres per capita. This means that for each individual living in the United Arab Emirates, over 26 acres are necessary to provide the consumptive and disposal needs for that person. By comparison, the footprint for the United States is 19.8 acres. Two additional questions that might be asked are the following: (i) is the footprint increasing with time and (ii) how does this footprint compare to the available capacity? Growth in the ecological footprint can be attributed to an increase in population, an increase in consumption, or both. Of the Western European countries, Sweden, Belgium, Portugal, Spain, and Switzerland have increasing footprints, while Denmark and the Netherlands are making concerted efforts to reduce their footprints. The most striking result of this ecological footprint analysis is that if the entire world lived like the people of the United States, it would take over five planet Earths to support the present world population.

THE LIMITS TO GROWTH

In 1972, a team of MIT experts wrote a report titled "The limits to growth" and presented it to scientists, journalists, and others and shortly published it as a book. It was the first time that computer modeling was used to answer the question as to whether the population would outgrow the planet and the resources available. The purpose of the study was to show the interrelationship between global growth factors like population, resources, persistent pollution, food production, and industrial activity. Based on this study, they predicted that if human beings continued to consume more than the environment was capable of providing, there would be an economic collapse and a sharp decline in population by 2030, which is not too far away.

This topic of overshoot and collapse was addressed again in *The Limits to Growth: The 30-Year Update* [6], which stated that "overshoot can lead to two different outcomes. One is a crash of some kind. Another is a deliberate turnaround, a correction, a careful easing down. We explore these two possibilities as they apply to human society and the planet that supports it. We believe that a correction is possible and that it could lead to a desirable, sustainable, sufficient future for all the world's peoples. We also believe that if a profound correction is not made soon, a crash of some sort is certain. And it will occur within the lifetimes of many who are alive today."

Although the 1972 report seemed to focus on a very negative scenario, it looked at various changes that could avert a collapse. One positive variable was looking at technological changes that increased agricultural productivity, reduced pollution, and provided an increase in the available supply of natural resources. Technological advancements would have a positive impact, but this alone could not avert a collapse. Social and cultural changes would also be necessary to reduce consumption and stabilize population growth. Since it had been 40 years since the report, data were collected and compared with the predictions. To mark the 40th anniversary of the report, experts gathered to discuss the challenges ahead into a sustainable future. Their concern was depicted in Figure 1.1 [7], which shows that the world is following the predictions of the study.

You can see that with 30 years of data, pollution, industrial output, population, and services per capita are all increasing as expected. At the same time, the remaining nonrenewable resources are decreasing a little slower (good), but food per capita is increasing a little faster than expected (bad).

The study was also concerned with sustainable development, which was defined by the notion that the developed nations can keep what they have while the poor people try to catch up, or, perhaps, keep on doing what we are all doing, but through technological advances we can expect less pollution and use fewer resources. Unfortunately, we are not succeeding with this expectation. We are currently consuming 50% more than what the Earth is able to provide, as explained earlier by the ecological footprint.

What we are consuming can be described as the different forms of industrial capital. This capital really refers to the machines and factories that produce the manufactured goods. These products manufactured by the industrial

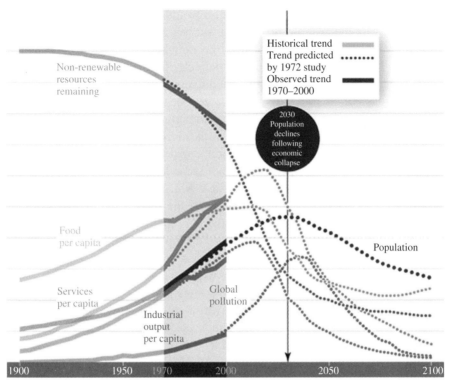

FIGURE 1.1   Tracking predicted trends.

capital can be defined as industrial output. This industrial output derived from industrial capital can be used to generate service capital for the service industry, like banks, schools, and hospitals, which provides services for the people. Industrial capital is also converted to agricultural capital to generate agricultural output. Likewise, it is converted to resource-obtaining capital to generate resource output. In addition, industrial capital is used to manufacture consumer goods. As each of these industrial outputs continues to grow, there is a need for more capital to be invested in the factories and machines that serve each of the outputs. Consequently, there may be an exponentially growing requirement for industrial output to expand the capacity for production in the future. This leads to more and more consumption.

CONSUMPTION FACTOR

Another way of looking at overuse of the Earth's resources is to talk about consumption. There is a great variation today in consumption of many nations in the world. Consumption is defined as the needs of people for survival in terms of food, energy, materials, and the disposal of waste. The disparity in the consumption rate is that it is 32 times greater in the United States, Canada, Western Europe, Japan, and Australia than in the developing countries [8].

Today, the world population is estimated to be around seven billion people, of whom only about one billion live in the fully developed countries listed earlier. By the middle of this century, it is estimated that the world population could grow to nine billion people, and there are questions as to whether the Earth can support this number of people, or will it collapse. It is not really a question of how many people are on this Earth, but what is the consumption rate of these people?

People in third-world countries are aware of a major difference in the consumption rate per capita, although they probably don't know the magnitude of the difference. In general, their goal is to catch up to the developed countries, but if they believe their chances of catching up are hopeless, they could get frustrated, angry, or even participate in terrorist activities. Another option is to emigrate to a first-world country like the United States and Western Europe, but then they would contribute to the consumption rate of that country.

If one considers the fastest growing economy in China, these people are already aspiring to increase their consumption rate to the 32 factor. If the Chinese were to succeed, it would be equivalent of doubling the world's consumption rate. If India were to do the same thing, the consumption rate would then triple. If the entire world had the same consumption rate as these first-world countries, it would be the same as having 72 billion people on this planet at the current consumption rates—and there is no way the Earth could handle this.

Since we are in no position to restrict the rest of world from improving their quality of life, the only answer is that the high-consuming countries mentioned earlier must lower their consumption rate. But will they do it for the benefit of the rest of the world? Whether they want to or not, they must reduce their consumption rate because what they are doing today is not sustainable.

If these countries reluctantly agree to reduce their consumption rate, does it mean that they will have to reduce their quality of life? Definitely not! For example, the people in Western Europe consume half as much oil (gasoline) per capita than the people in the United States. But the Western European standard of living is considered higher than that in the United States in terms of life expectancy, healthcare, infant mortality, vacation time, quality of public schools, and several other criteria. Does a large gas-guzzling automobile really contribute positively to any of these quality of life factors? Probably not!

The current state of the environment can also be presented by looking at four major environmental issues: (i) water scarcity, (ii) energy sources, (iii) climate change, and (iv) population growth.

CONSERVATION OF WATER

Water is a natural resource with a finite quantity. The amount of water on this planet 2000 years ago is the same as it is today, but the population during this time interval has gone from approximately 150 million to over 7 billion. But of all the water on the Earth, how much is readily available to all of the living creatures? Figure 1.2 [9] provides a summary of the current situation.

The Earth's surface is about 71% water; however, 97% of all the water on Earth is saline. Of the remaining 3%, 68.7% is in the form of ice caps and glaciers, 30.1% is groundwater, and 0.9% is in some other unavailable form. This leaves only 0.3% of the freshwater on Earth available to us on the surface, with 87% in lakes, 11% in swamps, and 2% in rivers. This means that only 0.1% of all the water on the Earth is available for industrial, agricultural, and human use. And of these three general uses, 70% is for agricultural use, 20% for industrial use, and only 10% for human consumption. Going further with the calculations results, only 0.01% of all the water on the Earth is being consumed by humans, and as the population grows, that leaves less for everyone.

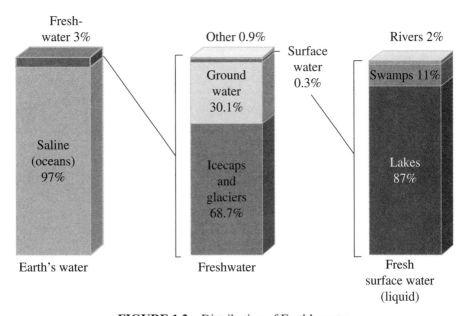

FIGURE 1.2   Distribution of Earth's water.

According to the United Nations, two-thirds of the world's population is projected to face water scarcity by 2025. In the United States, a federal report shows that 36 states are anticipating water shortages by 2013. In 2008, the state of Georgia tried, unsuccessfully, to move the state's border north in order to claim part of the Tennessee River.

The concern for this water shortage is partly due to the companies that require so much for their processes. It takes roughly 20 gallons of water to make a pint of beer, about 130 gallons of water to make a 2 liter bottle of pop, and 500 gallons of water to make a pair of Levi's stonewashed jeans. Why so much? For the pop, it includes the water used to grow the ingredients such as sugarcane. For the jeans, it includes the water used to grow, dye, and process the cotton.

Companies are now calculating the "water footprint" in order to manage better the water consumption. This is not dissimilar to the carbon footprint that organizations and individuals have been calculating for some years. The water-footprint concept was first developed in 2002 by A.Y. Hoekstra at the University of Twente in the Netherlands [10]. Following the water-footprint concept, studies were conducted to calculate the embedded, or virtual, water required for a product, which was then added to what is consumed directly. Embedded water includes everything from raising beef in South America, growing oranges in Spain, or growing cotton in Asia. By calculating the embedded water, you would learn that a typical hamburger takes 630 gallons of water to produce. Most of the water is used to grow the grain to feed the cattle. This represents more than three times the amount the average American uses every day for drinking, bathing, washing dishes, and flushing toilets.

At first glance, these large numbers representing water footprints for certain products seem very alarming. However, they are not necessarily bad if there is available water and it is well managed. Since most of the water is used for crops, it becomes part of the water cycle where it is eventually evaporated or it is run off. This water becomes temporarily unavailable for other uses, but that is not really a problem in an area that has plentiful water. If it doesn't return to the same aquifer or it returns as rainfall in another region, this could be a problem.

THE DEPLETION OF FOSSIL FUELS

In 1956, M. King Hubbert, a scientist with Shell Oil, proposed that fossil fuel production in a given region over time would follow a roughly bell-shaped curve without giving a precise formula [11]. Hubbert assumed that after fossil fuel reserves are discovered, production at first increases

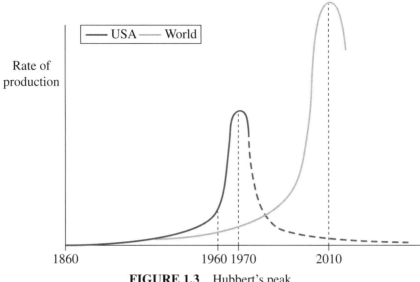

FIGURE 1.3 Hubbert's peak.

approximately exponentially, as more extraction commences and more efficient facilities are installed. At some point, a peak output is reached, and production begins declining until it approximates an exponential decline as shown in Figure 1.3.

The Hubbert curve suggests that the oil production rate increases as more reserves are discovered, and the rate peaks when half the estimated ultimately recoverable oil is produced. This is followed by a falling production rate, all along a classic bell curve. This same analysis has shown that it took 110 years to produce about 225 billion barrels of crude oil in the United States, but half of that oil was produced in the first 100 years and the second half in the next 10 years. In the United States, Hubbert predicted that this production rate peak would be achieved in 1970, the year when half of the estimated ultimately recoverable oil was utilized, and then production would start its steady decline. That, in fact, is what happened when the United States lost its preeminence as the world's leading producer of oil and caused a spike in gasoline prices and long lines at the pumps. On a global basis, this milestone was expected to occur around the year 2010, but Hubbert was not far off as US gasoline prices were highest during the summer of 2008.

According to this analysis, the current supply of fossil fuels, and oil in particular, is not only finite but decreasing rapidly. There may, however, be undiscovered reserves that could have an impact on this analysis. In any event, the quantities of oil, gas, and coal in the Earth's subsurface are finite, and their rate of consumption must be reduced.

CLIMATE CHANGE

A report from the United Nations Intergovernmental Panel on Climate Change (IPCC) stated that the Earth cannot tolerate more than a 3–5 °F increase in temperature. In order not to exceed this level, the carbon dioxide emissions must be reduced 60–80% of the 1990 levels by the year 2050. If you take into account the increase in population by 2050 and the corresponding increase in energy demand, to achieve this reduction is almost incomprehensible.

In 1988, the United Nations General Assembly created the IPCC with the task of reviewing and assessing the most recent scientific, technical, and socioeconomic information produced worldwide relevant to the understanding of climate change. Further, it would provide the world with a clear scientific view on the then current state of climate change and its potential environmental and socioeconomic consequences, notably the risk of climate change caused by human activity.

The first assessment report of the IPCC was presented in 1990 and along with subsequent reports led scientists to conclude that the Earth cannot tolerate more than a 3–5 °F increase in temperature. In order not to exceed this level, the carbon dioxide emissions must be reduced 60–80% of the 1990 levels by the year 2050. The first in a series of international meetings took place in 1992 in Rio de Janeiro, Brazil, called the Rio Earth Summit. As a result of that meeting, five years later, the Kyoto Protocol was adopted. It recognized that climate change was a result of greenhouse gases (GHG) created by human industrial activity. The idea was that rich nations, which had already benefited from industrialization, would reduce their GHG emissions in the first part of the treaty and developing nations would join in later. Milestones were created in various intervals through the year 2050. One of the milestones was to reduce GHG by 5% below 1990 levels by 2012. Instead, the world increased GHG by 58% above 1990 levels as the Kyoto Protocol came to an end.

At the next international meeting, which took place in Doha, Qatar, at the end of 2012, the developing countries once again demanded, as they did in Kyoto in 1997, for the rich countries to make a commitment to set real targets for reducing their GHG output. The rich nations then agreed to make some commitment toward a stronger legal agreement by 2015.

There are some scientists, while in the minority, who believe that global warming may exist and/or it is not anthropological. Regardless, reducing carbon dioxide emissions is like an insurance policy. If one assumes that global warming does exist and therefore takes the necessary action, the downside risk is minimal. If we learn in 20 years that global warming never really existed, it would have resulted in unnecessary development of renewable energy sources and possible introduction of a carbon tax. If, on the other hand, one assumes that global warming does not exist and therefore takes no action at all, what would happen if this assumption is eventually determined to be incorrect? The result on the world population could be catastrophic, with rising sea levels leading to flooding and droughts leading to dwindling food production.

POPULATION GROWTH

Each of the environmental issues described earlier, consumption, fossil fuel reserves, water scarcity, and climate change, are all related to the world population. Figure 1.4 [12] provides a summary of the historical as well as the projected growth in population.

The world population reached one billion people in 1804, two billion in 1927, five billion in 1987, six billion in 1999, and seven billion in 2011 [13]. As can be seen, the growth in population has been accelerating and is currently at an addition of 10 million people every six weeks. Most of this growth is in the developing countries, which may not be a major problem because of the lower consumption rate. However, as indicated earlier, some of these economies are increasing as are their consumption rates. China and India are examples of such growing economies.

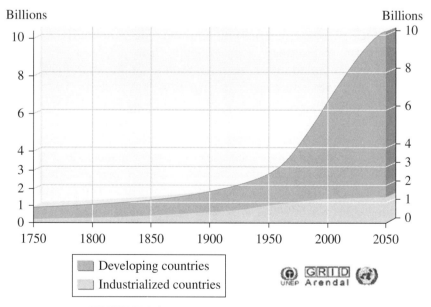

FIGURE 1.4   World population development.

THE ENVIRONMENT'S BIG FOUR

Today, these are the four major environmental concerns in the world, specifically:

- Water quality and quantity
- Depletion of fossil fuels
- Climate change resulting primarily from fossil fuels
- Population growth—eventually exceeding the Earth's capacity

Mitigating the impact of these four major environmental issues leads to an urgency for sustainable development.

REFERENCES

1. Brower D. *Let the Mountains Talk, Let the Rivers Run: A Call to Those Who Would Save the Earth*. San Francisco, CA: HarperCollins; 1995.
2. The Holy Bible, *Genesis* 1: 1–31.
3. Moyers B. *Earth on Edge*. Washington, DC: World Resources Institute and Public Affairs Television, Inc.; 2002.
4. Global Footprint Network. 2011. National Footprint Accounts. Available at www.footprintnetwork.org. Accessed 2013 Jul 19.
5. Venetoulis J, Talberth J. *Ecological Footprint of Nations 2005 Update*. Oakland, CA: Redefining Progress.
6. Meadows D, Randers J, Meadows D. *The Limits to Growth: The 30-Year Update*. White River Junction, VT: Chelsea Green Publishing Company; 2004.
7. Strauss M. Got corn? *Smithsonian Magazine*, p. 25, April 2012.
8. Diamond J. What's your consumption factor? *New York Times*, January 2, 2008.
9. US Geological Survey. Available at http://en.wikipedia.org/wiki/Water_distribution_on_Earth. Accessed 2013 Jul 19.
10. Hoekstra AY, editor. Virtual water trade. Proceedings of the International Expert Meeting on Virtual Water Trade; 2003 Feb; The Netherlands: IHE Delft.
11. Hubbert MK. Nuclear energy and the fossil fuels. Presented before the Spring Meeting of the Southern District, American Petroleum Institute, San Antonio, TX; 1956 Mar 7–9.
12. GRID-Arendal. Available at http://www.grida.no/graphicslib/detail/world-population-development_29db. Accessed 2013 Jun 28.
13. US Census Bureau. Available at http://www.worldometers.info/world-population/. Accessed 2013 Jul 19.

■■■■■■■ **CHAPTER 2**

Development of the Sustainability Concept and CSR

In 1983, the United Nations convened the World Commission on Environment and Development and was chaired by Dr. Gro Harlem Brundtland, who was the Prime Minister of Norway at the time. The agenda of this commission was the concern with the accelerating deterioration of the human environment and natural resources and the consequences of that deterioration for economic and social development. The final report was published in 1987 as *Our Common Future* [1] and resulted in the definition of sustainable development, or sustainability for short (p. 41):

> Sustainable development is development that meets the needs of the present without compromising the ability of future generations to meet their own needs.

This has become now well known as the Brundtland definition of sustainability, but there are many other definitions that really all mean the same— protecting the environment so it will be available for the people forever. A definition developed for the Center for Sustainable Enterprise at the Illinois Institute of Technology Stuart School of Business [2] is:

> The practice of sustainable development assures that the natural resources and energy we use to provide today's products and services will not deny future generations the resources necessary to meet their needs while building and preserving communities that are economically, socially and environmentally healthy.

The impact on the environment can also be presented as a mathematical relationship. For instance,

$$EB = P \times A \times T_1$$

Practical Sustainability Strategies: How to Gain a Competitive Advantage,
First Edition. Nikos Avlonas and George P. Nassos.
© 2014 John Wiley & Sons, Inc. Published 2014 by John Wiley & Sons, Inc.

where EB=environmental burden
  P=population
  A=affluence
  T_1=technology

Environmental burden refers to the negative impact on the environment, which is a function of (i) the population, (ii) affluence or consumption by this population, and (iii) technology that enhances or mitigates this impact. This formula can be restated as follows:

$$EB = \frac{P \times A}{T_2}$$

where T_2=the inverse of technology

But now, the technology is what creates products and services sustainably. Population and affluence are really inversely proportional: as the population increases, consumption tends to decrease. As the people become more educated and consumption increases, the population tends to decrease. However, if the population was to double and the consumption was to, say, increase by 10-fold, technology would have to increase by 20-fold to keep the environmental burden constant.

The bottom line to this discussion is that there is an urgent need for corporations, government agencies, nongovernmental organizations (NGOs), and individuals to find ways to be sustainable and to implement them. The following chapters in Part 2 will provide strategies for everyone, but for corporations in particular, to develop, maintain, or extend a competitive advantage *without* having a negative impact on the environment. Being more efficient in energy usage, water usage, and natural resource usage, and creating less waste are all very, very important. These can all be classified as being less bad. However, being *less bad is not enough.*

While most corporations are very concerned with profitability, implementing sustainability does not necessarily mean an opposition to profitability. In fact, these two terms are complementary. By striving for corporate sustainability, the company will achieve long-term profitability. This is the goal for all *corporate stakeholders*, a term first used by multinational companies in the late 1960s and early 1970s, referring to all people and entities upon which the company has an impact. At the same time, the term corporate social responsibility (CSR) was used to describe the process with an aim to embrace responsibility for the company's actions and encourage a positive impact through its activities on the environment, consumers, employees, communities, stakeholders, and anyone else in the public sphere who may also be considered as stakeholders. While this term was not widely used during the next three decades, it is currently used often to describe a company's integration of sustainability in its operations. Perhaps a more accurate name for the process would be corporate social responsibility and sustainability, or CSR and sustainability as used by McDonald's Corporation (Private conversation with Bob Langert, Sr. V.P. *CSR and Sustainability*. Oak Brook, IL: McDonald's Corporation).

In 2011, the European Commission defined CSR as "a concept whereby companies integrate social and environmental concerns in their business operations and in their interaction with their stakeholders on a voluntary basis" (Communication from the Commission to the European Parliament (A renewed EU strategy 2011–2014 for Corporate Social Responsibility)).

Corporate social responsibility concerns actions by companies over and above their legal obligations towards society and the environment. Certain regulatory measures create an environment more conducive to enterprises voluntarily meeting their social responsibility.

According to European Union (EU) policy, a strategic approach to CSR is increasingly important to the competitiveness of enterprises. It can bring benefits in terms of risk management, cost savings, access to capital, customer relationships, human resource management, and innovation capacity.

Because CSR requires engagement with internal and external stakeholders, it enables enterprises to anticipate better and take advantage of fast-changing societal expectations and operating conditions. It can therefore drive the development of new markets and create opportunities for growth.

By addressing their social responsibility, enterprises can build long-term employee, consumer, and citizen trust as a basis for sustainable business models. Higher levels of trust in turn help to create an environment in which enterprises can innovate and grow. Through CSR, enterprises can significantly contribute to the EU's treaty objectives of sustainable development and a highly competitive social market economy. CSR underpins the objectives of the Europe 2020 strategy for smart, sustainable, and inclusive growth, including the 75% employment target. Responsible business conduct is especially important when private sector operators provide public services. Helping to mitigate the social effects of an economic crisis, including job losses, is part of the social responsibility of enterprises. CSR offers a set of values on which to build a more cohesive society and on which to base the transition to a sustainable economic system.

CSR TODAY: FROM SHAREHOLDER VALUE TO STAKEHOLDER VALUE

Corporations today are being met by greater pressure from consumer and stakeholder groups to act upon societal and environment obligations. In essence, their actions are to mitigate the negative impact their activities are having upon the wider world. More and more, business leaders are themselves coming to the realization that their activities reach beyond the sphere of the business marketplace, impacting social, environmental, and political realms as well, and often negatively, and have implemented various measures, such as voluntary CSR reporting and recycling programs, to counteract the social and environmental tolls of global business.

With the rise of consumer education and activism, global social consciousness has become a hot topic of discussion, with business leaders jumping at the chance to exert themselves as champions of the people and promote their businesses as "green" or "socially responsible." And in the wake of a continually evolving set of ethical business guidelines comes the redefinition of the very concept of *business*. No longer is the "shareholder value prospect," that is, maximum profit, the only player in the game, but its rival, CSR, a largely stakeholder and consumer concern, has emerged as the new Most Valuable Priority (MVP). This reshuffle of priorities reflects the ability of globally responsible and creative business leaders to remain profitable while expending company resources towards the betterment of society, the transparency of business, and the protection of the environment.

Stakeholders—customers, employees of all ranks, suppliers, partners, affiliated institutes and governments, shareholders, and "owners"—represent the most fertile and diverse source of easily procured information available to a business or an organization. It should, therefore, come as no surprise that stakeholder engagement and dialog is crucial for the profitability and continued existence of a company, as stakeholder opinions can be utilized to guide its vision and mission and reflect on its performance. To simultaneously realize the potential of stakeholder groups and to retain their valuable insights, a company must be willing to engage stakeholders in frequent dialog, weigh their opinions, and allow them to participate in various decision-making processes. If this is not done, the company runs the risk of losing stakeholder capital—whether this is monetary, intellectual, or labor capital—which could prove potentially catastrophic to the life of the company in the long term.

Follow-up and feedback with stakeholders provide a cycle of opportunity for the growth and improvement of a company. Diverse stakeholder groups, when engaged in frequent dialog, can steer a company in a profitable and productive direction through the provision of perspectives on a variety of issues, from within a range of industries, geographic locals, demographic groups, and cultural backgrounds. For efficient and effective stakeholder dialog to be established, constant, clarifying communication between parties is essential, such as the identification of common goals among stakeholder groups and the company, as is the engagement and participation of stakeholders in relevant fora. Building relationships that transcend traditional business boundaries is a proven way to involve stakeholders in the mission of the company on a personal and emotional level and evokes a genuine desire to learn from partnerships.

CSR MEASURING AND REPORTING

In addition to CSR, another term being used by business community is environmental, social, and corporate governance (ESG). This refers to the three main areas of concern that have developed as the central factors in measuring the sustainability and ethical impact of an investment in a company or business. Whether a company is concerned about CSR or ESG, how does it know whether it is meeting its goals and objectives? And how does the investment community know which companies are achieving the CSR and ESG objectives?

The best way to really understand if your organization is meeting any goals and objectives is to take appropriate measurements of the key performance indicators (KPI). But it is important to measure both the environmental/social indicators and the financial indicators because a balanced scorecard is important. In other words, the balanced scorecard consists of financial and nonfinancial indicators.

Another approach at obtaining this balance is to consider the environmental, social, and economic aspects of the company, or the triple bottom line (TBL) as coined by John Elkington [3]. It is also known as planet, people, and profit (P^3) or earth, equity, and economy (E^3). This is often shown as a triangle (Fig. 2.1) depicting a three-legged chair as all three legs are

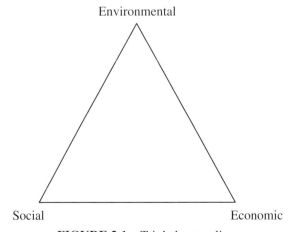

FIGURE 2.1   Triple bottom line.

FIGURE 2.2   Realistic triple bottom line.

necessary. A more realistic depiction can be seen in Figure 2.2 that shows the planet and environment as a large circle. Within the environment, in a smaller circle, are the people that depend on the environment. And what is dependent on the environment and society is the economy. But there needs to be a measurement to understand achieving the TBL. The Global Reporting Initiative (GRI), developed in 1997, is the internationally accepted standard for TBL reporting. The GRI was created to bring consistency to the TBL reporting process by enhancing the quality, thoroughness, and value of sustainability reporting.

The following chapters in Part 3 will provide readers with various measuring standards as well as reporting standards. The benefits to the organization will also be outlined.

THE SUSTAINABLE DEVELOPMENT CONCEPT THOUSANDS OF YEARS AGO

Although the term CSR is used today, concepts analogous to social responsibility have existed for centuries. The idea that those in an advantageous position, either in terms of money or power, should conduct in a socially responsible way can be traced back to classical Athens of 500 B.C.

This view supports that from the early days of civilization, it was understood, in some cases by the state or in others by wealthy individuals of the time, that it would only be fairer for the rest of society if such individuals shared a portion of their amassed wealth with their fellow citizens. This reality came into being from the realization of the fact that had it not been for the rest of society, wealth making would probably be an impossibility. Thus, bearing this realization in mind, certain laws were voted by the state which at the time reflected precisely and directly the opinion of the people. Namely, these were the liturgies, "a wide range of public service paid for out of the pockets of the people owning substantial property" [4]. There were two different kinds of liturgies, the regular liturgies that had a religious character and the extraordinary liturgies that had a military character.

Aristotle's "magnificent man," who bestows collective benefits upon his community out of his own wealth, is depicting the exact definition of "euergetism" in ancient Greece. The nobles, who because of their birth and wealth controlled access to all essential services, were expected to provide various services to their cities in exchange for the public bestowal of honor from the inhabitants. In the Art of Rhetoric, Aristotle states that true wealth consists in doing good, that is, in monetary handouts, giving of scarce and costly gifts, and helping others to maintain an existence [5].

A last notion connecting CSR and ancient Greece is that of heroism. Nowadays the term "hero" has many different meanings. It addresses to people whose work is extremely risky like firemen or to people whose work is challenging like teachers. "Hero" originates from the Greek word hêrôs. Although in ancient Greece this word was referring only to warriors, later on it was introduced for the people who have offered in favor to the local community but they are now dead. In order for someone to obtain the heroic status, he or she should deal with local sponsorship. There were no exact rules that someone had to follow in order to attain a heroic status. Someone who was involved in the matters of public interest and his or her goal was the welfare of the community could be characterized as a hero.

REFERENCES

1. World Commission on Environment and Development (WCED). Available at http://en.wikipedia.org/wiki/Brundtland_Commission. Accessed 2013 Jun 29, later published as *Our Common Future*. Oxford: Oxford University Press; 1987.
2. Kusz JP, Nassos GP. *Roundtable Discussion to Introduce the Center for Sustainable Enterprise*. Chicago, IL: Illinois Institute of Technology; July 2000.
3. Elkington J. *Cannibals With Forks: The Triple Bottom Line of 21st Century Business*. Oxford: Capstone Publishers; 1997.
4. Andrews A. *Greek Society*. London: Penguin Group; 1967.
5. Joint Association of Classical Teachers. *The World of Athens: An Introduction to Classical Athenian Culture*. Cambridge: Cambridge University Press; 1992.

PART 2

SUSTAINABILITY STRATEGIES

CHAPTER 3

The CERES and Other Principles

It is important to note that there are many activities that an organization can undertake for the benefit of the environment. Some of these activities may be sustainable, but many of them are not truly sustainable yet very beneficial as they reduce the negative impact on the environment. These environmentally beneficial activities embraced by corporations become the guidelines that are followed, better known as "principles."

As a result of the Exxon Valdez oil spill in 1989, a group of investors launched the Coalition for Environmentally Responsible Economies (CERES). This group, now just called CERES, consists of over 130 institutional and socially responsible investors (SRIs), environmental and social advocacy groups, and other public interest organizations. The Coalition works to promote sustainability by moving companies, policymakers, and other market players to incorporate environmental and social challenges into their decision-making and improve corporate practices to create a more just and prosperous global economy [1]. A list of the CERES members can be seen at http://www.ceres.org/about-us/coalition/coalition-members. These members include environmental and social nonprofit groups such as NRDC, Union of Concerned Scientists and Oxfam, institutional investors such as the California and New York public pension funds, SRIs, labor unions, and other key stakeholders. In recent years, CERES has expanded the coalition's expertise to include more social issues such as diversity and human rights.

Shortly after its creation, CERES adopted a 10-point code of corporate environmental conduct to promote environmental awareness and accountability. Companies that adopted the CERES Principles were on their way to become environmentally conscious and eventually more sustainable.

Practical Sustainability Strategies: How to Gain a Competitive Advantage,
First Edition. Nikos Avlonas and George P. Nassos.
© 2014 John Wiley & Sons, Inc. Published 2014 by John Wiley & Sons, Inc.

CERES PRINCIPLES

Each company that adopts the CERES Principles must commit to all 10 principles [2]:

1. *Protection of the Biosphere*   We will reduce and make continual progress toward eliminating the release of any substance that may cause environmental damage to the air, water, or the earth or its inhabitants. We will safeguard all habitats affected by our operations and will protect open spaces and wilderness while preserving biodiversity.

2. *Sustainable Use of Natural Resources*   We will make sustainable use of renewable natural resources, such as water, soils, seafood, plants, and forests. We will conserve nonrenewable natural resources through efficient use and careful planning.

3. *Reduction and Disposal of Wastes*   We will reduce and where possible eliminate waste through source reduction and recycling. All waste will be handled and disposed of through safe and responsible methods. The reuse of the waste material will also be considered.

4. *Energy Conservation*   We will conserve energy and improve the energy efficiency of our internal operations and of the goods and services we sell. We will make every effort to use environmentally safe and sustainable energy sources such as solar, wind, and geothermal.

5. *Risk Reduction*   We will strive to minimize the environmental, health, and safety risks to our employees and the communities in which we operate through safe technologies, facilities, and operating procedures, and by being prepared for emergencies.

6. *Safe Products and Services*   We will reduce and where possible eliminate the use, manufacture, or sale of products and services that cause environmental damage or health or safety hazards. We will inform our customers of the environmental impacts of our products or services and try to correct unsafe use.

7. *Environmental Restoration*   We will promptly and responsibly correct conditions we have caused that endanger health, safety, or the environment. To the extent feasible, we will redress injuries we have caused to persons or damage we have caused to the environment and will restore the environment.

8. *Informing the Public*   We will inform in a timely manner everyone who may be affected by conditions caused by our company, which might endanger health, safety, or the environment. We will regularly seek advice and counsel through dialog with persons in communities

near our facilities. We will not take any action against employees for reporting dangerous incidents or conditions to management or to appropriate authorities.

9. *Management Commitment* We will implement these principles and sustain a process that ensures that the Board of Directors and Chief Executive Officer are fully informed about pertinent environmental issues and are fully responsible for environmental policy. In selecting our Board of Directors, we will consider demonstrated environmental commitment as a factor.

10. *Audits and Reports* We will conduct an annual self-evaluation of our progress in implementing these principles. We will support the timely creation of generally accepted environmental audit procedures. We will annually complete the CERES Report, which will be made available to the public.

If you carefully read the content of each of these 10 principles, you will note that each one was prepared for the benefit of protecting the environment. These principles do not necessarily make a company sustainable, but they may be considered a precursor to sustainability. Nevertheless, many of top companies in the world have adopted the CERES Principles as seen in Table 3.1.

About 20 years after its inception, CERES went to the next step and created a roadmap on achieving sustainability and in 2010 published *The 21st Century Corporation: The CERES Roadmap for Sustainability*. It analyzes the drivers, risks, and opportunities involved in making the shift to sustainability, and details strategies and results from companies that are taking on these challenges. More robust and comprehensive than the CERES Principles and updated for the twenty-first century, the *CERES Roadmap* is designed to provide a comprehensive platform for sustainable business strategy and for accelerating best practices and performance.

The Roadmap sets out 20 expectations for sustainability that companies should start implementing now to be considered sustainable going forward. It is raising the bar for leadership. These expectations are laid out in four broad areas that are key for corporate sustainability: governance, stakeholder engagement, disclosure, and performance. All of the expectations presented in the Roadmap need to be addressed for a company to achieve a comprehensive and coherent sustainable business strategy. The full report has more than 200 company best practice examples across 20 sectors. Many companies have started this journey, from heavy industry to consumer products, and the Roadmap includes a full range of examples to demonstrate what is possible now and where companies need to go in the future. The report features more than 250 resources and tools from a wide range of global experts, organizations, and thought leaders [4].

TABLE 3.1 CERES Network Members [3]

Advanced Micro Devices	Dunkin' Brands	Northeast Utilities
Allstate Insurance	Earth Color	PepsiCo
Anvil Knitwear	eBay	PG&E, Corp.
APS	Eileen Fisher	Promotional Product Solutions
Aspen Skiing Company	EMC	Recyclebank
Aveda	Energy Management, Inc.	Recycled Paper Printing
Bank of America	Exelon	Saunders Hotel Group
Baxter International	Ford Motor Company	Seventh Generation
Ben & Jerry's	Gap, Inc.	Sodexo
Best Buy	General Mills	Sprint
Bloomberg	General Motors	State Street, Corp.
Brighter Planet	Green Mountain Coffee	Suncor Energy
Brown-Forman, Corp.	Green Mountain Energy	Sunoco, Inc.
CA Technologies	Green Mountain Power	SustainAbility
Carbon Credit, Corp.	Haley & Aldrich	The Co-operators Group
Catholic Healthcare West	Intuit	The Coca-Cola Company
Citi	ITT, Corp.	The North Face
Clif Bar & Company	Jones Lang LaSalle	The Walt Disney Company
Concept A	Legg Mason	Timberland
Cone	Levi Strauss & Co.	Time Warner, Inc.
Consolidated Edison	McDonald's, Corp.	Vancity Savings Credit Union
credit360	National Grid plc	Virgin America
Curtis Packaging	NativeEnergy	William McDonough Partners
Dell, Inc.	Nike	YSI, Inc.

The Roadmap refers to four key drivers of sustainability. They are (i) competition for natural resources, (ii) climate change, (iii) economic globalization, and (iv) connectivity and communications. It also lists the 20 expectations for sustainability, which are summarized in the succeeding text, but a thorough description of this Roadmap can be seen on the CERES website [5].

ROADMAP EXPECTATIONS

Governance for Sustainability

G1: Board Oversight   The Board of Directors will provide oversight and accountability for corporate sustainability strategy and performance. A committee of the board will assume specific responsibility for sustainability oversight within its charter.

G2: Management Accountability   The CEO and company management—from C-suite executives to business unit and functional heads—will be responsible for achieving sustainability goals.

G3: Executive Compensation   Sustainability performance results are a core component of compensation packages and incentive plans for all executives.

G4: Corporate Policies and Management Systems   Companies will embed sustainability considerations into corporate policies and risk management systems to guide day-to-day decision-making.

G5: Public Policy   Companies will clearly state their position on relevant sustainability public policy issues. Any lobbying will be done transparently and in a manner consistent with sustainability commitments and strategies.

Stakeholder Engagement

S1: Focus Engagement Activity   Companies will systematically identify a diverse group of stakeholders and regularly engage with them on sustainability risks and opportunities, including materiality analysis.

S2: Substantive Stakeholder Dialog   Companies will engage stakeholders in a manner that is ongoing, in-depth, and timely, and involves all appropriate parts of the business. Companies will disclose how they are incorporating stakeholder input into corporate strategy and business decision-making.

S3: Investor Engagement   Companies will address specific sustainability risks and opportunities during annual meetings, analyst calls, and other investor communications.

S4: C-Level Engagement   Senior executives will participate in stakeholder engagement processes to inform strategy, risk management, and enterprise-wide decision-making.

Disclosure

D1: Standards for Disclosure   Companies will disclose all relevant sustainability information using the Global Reporting Initiative (GRI) guidelines as well as additional sector-relevant indicators.

D2: Disclosure in Financial Filings   Companies will disclose material sustainability issues in financial filings.

D3: Scope and Content   Companies will regularly disclose significant performance data and targets relating to their global direct operations, subsidiaries, joint ventures, products, and supply chain. Disclosure will be balanced, covering challenges as well as positive impacts.

D4: Vehicles for Disclosure   Companies will release sustainability information through a range of disclosure vehicles, including stand-alone reports, annual reports, financial filings, websites, and social media.

D5: Product Transparency Companies will provide verified and standardized sustainability performance information about their products at point of sale and through other publicly available channels.

D6: Verification and Assurance Companies will verify key sustainability performance data to ensure valid results and will have their disclosures reviewed by an independent, credible third party.

Performance

P1: Operations Companies will invest the necessary resources to achieve environmental neutrality and to demonstrate respect for human rights in their operations. Companies will measure and improve performance related to greenhouse gas (GHG) emissions, energy efficiency, facilities and buildings, water, waste, and human rights.

P2: Supply Chain Companies will invest the necessary resources to achieve environmental neutrality and to demonstrate respect for human rights in their operations. Companies will measure and improve performance related to GHG emissions, energy efficiency, facilities and buildings, water, waste, and human rights.

P3: Transportation and Logistics Companies will systematically minimize their sustainability impact by enhancing the resiliency of their logistics. Companies will prioritize low-impact transportation systems and modes, and address business travel and commuting.

P4: Products and Services Companies will design and deliver products and services that are aligned with sustainability goals by innovating business models, allocating R&D spend, designing for sustainability, communicating the impacts of products and services, reviewing marketing practices, and advancing strategic collaborations.

P5: Employees Companies will make sustainability considerations a core part of recruitment, compensation, and training and will encourage sustainable lifestyle choices.

Summary of Key Findings of Roadmap

In the CERES Report "The road to 2020: corporate progress on the road to sustainability" [6], it reported on the findings for each of the four areas of activities. The findings of the report represent a collaborative effort by CERES and Sustainalytics, a partnering firm, to assess the progress of the companies and to identify noteworthy trends and business practices. Its conclusion was that companies are moving, albeit too slowly, given the urgency of the sustainability challenges that are being faced.

With respect to each key expectation of the Roadmap evaluated, companies were placed in a performance tier:

- Tier 1—Setting the Pace
- Tier 2—Making Progress
- Tier 3—Getting on Track
- Tier 4—Starting Out

Some companies have made great strides; others are standing still; and most are somewhere in between. This means that for most companies, the opportunities to transform themselves are great.

In the area of Governance and Sustainability, 26% of the 600 companies (157 companies) including Alcoa, Xcel, and Intel are in Tiers 1 and 2 for their governance strategies on sustainability. More than half, however, are in Tier 4.

In Stakeholder Engagement, almost 24% of companies have some degree of meaningful stakeholder engagement, including Baxter and Ford, which demonstrates ongoing and long-term engagement with a diversity of stakeholders and discloses how they consider stakeholder feedback in business decision-making and strategy. However, nearly half of the companies assessed disclose no efforts on stakeholder engagement and would be classified in Tier 4.

In the Disclosure area, of the 600 companies, 49% (293 companies) are publishing sustainability reports, with 29% (176 companies) using the GRI guidelines. This still leaves almost half of the companies without a sustainability report—again in Tier 4.

In the Performance area of the Roadmap, this was divided into four subcategories. Under Reducing GHG Emissions, nearly half of the companies, 47%, are making some progress in reducing GHG emissions by reducing electricity demand, procuring renewable energy, and ramping up energy efficiency. A third of the 600 companies have in place time-bound targets for reducing GHG emissions for direct operations.

In Water Management, among four particularly water-intensive sectors analyzed, Food & Beverage, Footwear & Apparel, Oil & Gas, and Utilities, 25% of the companies, including Coca-Cola and Exelon, have undertaken assessments to identify specific water-related risks, such as geographic-specific exposure.

With respect to Human Rights performance, only 13% (80 companies) of the 600 companies evaluated on policies and programs are ranked in Tiers 1 and 2. Top-performing companies for this expectation include 3M, General Electric, and Hess, for which policies covering freedom of association, elimination of discrimination, human rights, and working conditions were evaluated.

And finally, under Supply Chain, nearly half—43% or 259 companies—of the 600 companies have a supplier code in place and nearly 10% (55 companies) make explicit reference to relevant International Labor Organization (ILO) conventions. Overall, only 25% of the 600 companies disclose some amount of supply chain monitoring and performance information, including Nike and Hewlett Packard.

CERES and Reporting

In 1997, a steering committee formed by CERES began on a project to create a disclosure process for sustainability [7]. This led to the GRI (mentioned earlier and covered in detail in Chapter 16) that is now used by over 1800 companies. CERES has also spearheaded dozens of breakthrough achievements with more than 80 companies, such as Best Buy launching a cutting-edge buyback program for its electronics products; Canadian oil sands producer Suncor cutting water use by 22% and committing to additional reductions and rapid reclamation of tailings ponds; Ford committing to a 30% carbon emissions reduction target by 2020; and Bank of America setting and meeting GHG reduction goals in utility lending [8].

OTHER PRINCIPLES FOR COMMERCE

While the CERES Principles are probably the most recognized and implemented principles for manufacturing and service corporations, there have been many other principles proposed as part of the "sustainability revolution" [9].

The Hannover Principles

The Hannover Principles were developed in 2000 by William McDonough for EXPO 2000, which took place in Hannover, Germany. The theme of this event was "Humanity, Nature and, Technology" incorporating the elements of ecological design. The principles are:

1. Insist on rights of humanity and nature to coexist.
2. Recognize interdependence.
3. Respect relationships between spirit and matter.
4. Accept responsibility for the consequences of design.
5. Create safe objects of long-term value.

6. Eliminate the concept of waste.
7. Rely on natural energy flows.
8. Understand the limitations of design.
9. Seek constant improvement by the sharing of knowledge.

The focus of these principles is the relationship between nature and design. More specifically, they were conceived "to inform the international design community of the issues inherent in the consideration of sustainable design, rather than to provide an ecological checklist of construction" [10].

The Precautionary Principle

The origin of the precautionary principle comes from one of Germany's basic principles for the protection of the environment. There was even a German law for "precaution" or "foresight" that states (9, p. 55):

> Environmental policy is not fully accomplished by warding off imminent hazards and the elimination of damage which has occurred. Precautionary environmental policy requires furthermore that natural resources are protected and demands on them are made with care.

This principle was subsequently mentioned in the Rio Declaration on the Environment in 1992 and defined in more detail in 1998 at Wingspread in Racine, Wisconsin, that later became known as the Wingspread Statement [11].

ADDITIONAL PRINCIPLES

In Andre Edwards' *The Sustainability Revolution: Portrait of a Paradigm Shift* [9], there is mention of many other principles depending on the focus. While the CERES Principles are, by far, the most prevalent in the business community, here is a list of others that could be considered.

Community

Principles of Sustainable Development for Minnesota
Ontario Round Table on Environment and Economy (ORTEE)
The Netherlands National Environmental Policy Plan (NEPP)
The Earth Charter
International Council for Local Environmental Initiatives (ICLEI)

Natural Resources

American Petroleum Institute (API) Environmental, Health, and Safety Principles

Forest Stewardship Council (FSC) Principles and Criteria for Forest Stewardship

Marine Stewardship Council (MSC) Principles and Criteria for Sustainable Fishing

The Asilomar Declaration for Sustainable Agriculture

Ecological Design

The Five Principles of Ecological Design

The Todds' Principles of Ecological Design

The Sanborn Principles

US Green Building Council and LEED

Biosphere

Deep Ecology's Basic Principles

Charter of Rights and Responsibilities for the Environment

Biomimicry Principles

Mollisonian Permaculture Principles

REFERENCES

1. Available at http://www.ceres.org/about-us/coalition. Accessed 2013 Jun 29.
2. Available at http://www.ceres.org/about-us/our-history/ceres-principles. Accessed 2013 Jun 29.
3. Available at http://www.ceres.org/company-network/company-directory. Accessed 2013 Jun 29.
4. Available at http://www.ceres.org/company-network/ceres-roadmap. Accessed 2013 Jun 29.
5. Available at http://www.ceres.org/resources/reports/ceres-roadmap-to-sustainability-2010. Accessed 2013 Jun 29.
6. Available at http://www.ceres.org/roadmap-assessment. Accessed 2013 Jun 29.
7. Available at https://www.globalreporting.org/information/about-gri/what-is-GRI/Pages/default.aspx. Accessed 2013 Jun 29.

8. Available at http://www.ceres.org/about-us/our-history. Accessed 2013 Jun 29.

9. Edwards AR. *The Sustainability Revolution: Portrait of a Paradigm Shift.* Gabriola Island, BC: New Society Publishers; 2005.

10. Available at http://www.mcdonough.com/wp-content/uploads/2013/03/Hannover-Principles-1992.pdf. Accessed 2013 Jul 19.

11. Available at http://www.sehn.org/state.html#w. Accessed 2013 Jun 29.

CHAPTER 4

The Natural Step

In Chapter 3, we discussed the development of various principles that can be considered as guides to achieving environmental and social sustainability. One of the principles not mentioned is the Natural Step. This one is somewhat unique and deserves special mention.

In the late 1980s, Karl-Henrik Robért, a Swedish medical doctor and cancer-treatment researcher, couldn't help but be confused between two groups of people. On one hand, there were some people that were more concerned with increasing their personal wealth than being concerned with the environment and its potential impact on children. And on the other hand, he was seeing concerned parents bringing to the hospital cancer-diseased children.

Dr. Robért became obsessed about the potential destruction of the environment. He saw humankind running into a funnel of declining life-sustaining resources and increasing demands [1]. He was aware of the various environmental issues like climate change, increased pollution, and water scarcity, all of which would be compounded with population growth. At the same time, he was not sure whether businesses would focus beyond short-term profits to long-term growth for the good of the environment, or whether there would be any legislation to make it happen. He then began to think: "What if we could use our agreement on the basic understanding of cells as a platform to understanding the requirements for the continuation and well-being of human life? From the perspective of that smallest unit of life, which merges us in a way that goes beyond politics and belief systems, we could build consensus among governments, business people, and environmentalists about what must at least be agreed to safe guard prosperous life" [2].

Practical Sustainability Strategies: How to Gain a Competitive Advantage,
First Edition. Nikos Avlonas and George P. Nassos.
© 2014 John Wiley & Sons, Inc. Published 2014 by John Wiley & Sons, Inc.

Dr. Robèrt drafted a first version of such a "consensus document" and sent this draft to a broad cross section of scientists, including over 50 ecologists, chemists, physicists, and medical doctors in Sweden and asked for their input. Twenty-one drafts later, there was at last consensus about what is in principle needed to sustain the human civilization on Earth. With the support of His Majesty the King of Sweden, Carl XVI Gustaf, this "consensus document" and accompanying audiotape was sent to every household and school in Sweden.

Based on this document, in 1989, Dr. Robért started the Natural Step, an environmental not-for-profit organization in Sweden. He wanted to address these troubling environmental problems and seek advice and expertise of a growing circle of scientists, economists, business leaders, and other stakeholders in society. This dialog with key participants spread throughout the world, and as a result, Natural Step organizations were launched in the United States and other countries like Canada, the United Kingdom, Australia, and New Zealand.

Robért and his advisors developed a simple framework to look at the impact of the environment on businesses and how to provide a means for businesses to become sustainable. The framework was based on systems thinking, that is, looking at the whole system, but in such a way that it can be made simple, thus allowing it to apply to almost any organization. The framework includes four core processes [1]:

- Perceiving the nature of the unsustainable direction of business and society and the self-interest implicit in shifting to a sustainable direction
- Understanding the first-order principles for sustainability, that is, the four system conditions
- Strategic visioning through "backcasting" from a desired sustainable future
- Identifying strategic steps to move the company from its current reality towards its desired vision

He then applied the funnel effect to show how the supply and demand of environmental resources are converging. This impact on the environment can be best described as follows [3]:

Global society is currently unsustainable. Because we only have one planet Earth, and the laws of thermodynamics are such that matter and energy cannot be created or destroyed, there are limits to how much we can grow—both in terms of sources (the resources we use) and *sinks* (the natural systems where we deposit our waste). We are currently surpassing those limits because, on

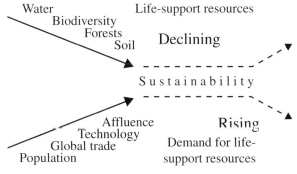

FIGURE 4.1 The funnel.

the one hand, population, consumption, competitiveness, and pollution are *systematically* increasing, while on the other hand, resources, purity, ecosystems, and social equity are *systematically* decreasing.

A "whole systems" perspective is necessary in sustainable development, as the various problems (greenhouse gases, extreme poverty, deforestation, illiteracy, etc.) are inherently interrelated and complex.

The funnel (Fig. 4.1) metaphor is a way of thinking about the unsustainable path that global society is on, where our space for deciding on options is becoming narrower and narrower per capita. This is very different from the illusion that limits to growth are represented by a "cylinder" where isolated social and ecological impacts can be addressed and "solved" separately.

The negative effects of society's unsustainable path can be described as "hitting the funnel walls." Organizations (communities, businesses, project teams, etc.) can feel the impact of hitting the funnel walls in a variety of ways, including:

- Increased costs for resources and waste management
- Lost investment in projects that quickly become obsolete
- Stricter legislation
- Litigation
- Loss of market share to more cutting edge
- Insurance costs
- Consumer and shareholder activism
- Loss of good reputation

Organizations that have an understanding of the funnel will be better able to act strategically, communicate more effectively to internal and external stakeholders, and lead the shift towards a sustainable society.

THE FOUR SYSTEM CONDITIONS FOR SUSTAINABILITY

In order to build this framework for sustainability, society must adopt the four system conditions in order to be ecologically sustainable. The Natural Step summarizes the four system conditions as follows: in order for a society to be sustainable, nature's functions and diversity are not systematically:

1. **Subject to increasing concentrations of substances extracted from the Earth's crust.**
2. **Subject to increasing concentrations of substances produced by society.**
3. **Impoverished by overharvesting or other forms of ecosystem manipulation.**
4. **Resources are used fairly and efficiently in order to meet basic human needs worldwide.**

SYSTEM CONDITION 1—SUBSTANCES FROM THE EARTH'S CRUST MUST NOT SYSTEMATICALLY INCREASE IN THE ECOSPHERE

This means that materials like fossil fuels, metals, minerals, and other natural resources must not be extracted from the Earth at a faster rate than they are reentering or depositing into the Earth's crust or even its atmosphere. A good example of violating this condition is the extraction of fossil fuels like oil or coal. When the fossil fuel is combusted, the carbon in the coal, for instance, is converted to carbon dioxide. For most of the human existence, the carbon dioxide emitted would eventually be reabsorbed by trees and other vegetation. The rate of emitting carbon dioxide was equal to the rate of absorption. Consequently, the carbon cycle was in balance and there was no violation of System Condition 1. The concentration of carbon dioxide was consistently about 280 parts per million. However, we are now emitting carbon dioxide at a faster rate than it is being absorbed, thus causing an increase in the carbon dioxide concentration in the atmosphere—now over 400 parts per million. This cover of carbon dioxide is preventing the release of reflected sun rays back into the atmosphere, thus causing an increase in the temperature of the Earth—global warming.

If this condition is not met, the concentrations of substances in the ecosphere will increase and eventually reach some limits beyond which it will be almost impossible to reverse. In the case of carbon dioxide, many scientists believe that it will be impossible to reverse, even if we stopped emitting carbon dioxide. In a report by the National Oceanic and Atmospheric

Administration, it stated that "changes in surface temperature, rainfall, and sea level are largely irreversible for more than a thousand years after carbon dioxide emissions are completely stopped." In an interview with the lead author, S. Solomon, she said, "People have imagined that if we stopped emitting carbon dioxide that the climate would be back to normal in 100 or 200 years. What we're showing here is that that's not right" [4].

SYSTEM CONDITION 2—SUBSTANCES PRODUCED BY SOCIETY MUST NOT SYSTEMATICALLY INCREASE IN THE ECOSPHERE

This means that substances must not be produced at a faster rate than the rate at which they can be broken down and integrated back into nature or redeposited into the Earth's crust. If this condition is not met, the concentration of substances in the ecosphere will continue to increase and eventually reach undesirable limits beyond which it will be difficult to reverse. In most cases, this upper limit is not known.

It is critical that we eliminate our contribution to the progressive buildup of chemicals and compounds produced by society such as dioxins, polychlorinated biphenyls (PCBs), and dichlorodiphenyltrichloroethane (DDT). In addition to these very toxic chemicals, there are many other substances produced by society that are accumulating in nature and are detrimental to the environment.

A very interesting example of the violation of System Condition 2 is that of ozone. The chemical name for ozone is trioxygen as it consists of three oxygen atoms as opposed to two atoms for the oxygen molecule. Electromagnetic radiation or the sparking from high-voltage applications will cause the oxygen molecule in the air to disassociate into two oxygen atoms, and then each atom will combine with another oxygen molecule to form ozone. Devices that require high voltages and as a result produce ozone are arc welders, ionic air purifiers, laser printers, and photocopiers. Ozone is also created by precursors such as nitrogen oxides, carbon monoxide, and various volatile organic compounds (VOCs).

Ozone in the lower atmosphere is a pollutant and may cause respiratory problems. In the upper atmosphere, however, ozone is beneficial as it prevents potentially damaging electromagnetic radiation from reaching the Earth's surface. Other man-made substances like chlorinated fluorocarbons (CFCs) break up in the atmosphere, and the free chlorine or fluorine atoms become catalysts to break up the ozone molecule. This has caused the formation of an "ozone hole" in the atmosphere that allows electromagnetic radiation to reach the Earth. So this is an example of the formation and destruction of ozone being the result of violating System Condition 2.

As a result of the damage created by the CFCs, the Montreal Protocol of 1987 called for the reduction and the eventual ban of CFC chemicals for use as refrigerants and other applications. The major chemical companies then started producing hydrogenated chlorinated fluorocarbons (HCFC), which are considerably less destructive to the ozone layer [5]. Electrolux, the refrigerator and appliance company, decided not to replace their CFC chemicals with a less toxic HCFC because this still went against System Condition 2. Instead, they opted to research for a totally biologically harmless substitute.

SYSTEM CONDITION 3—THE PHYSICAL BASIS FOR PRODUCTIVITY AND DIVERSITY OF NATURE MUST NOT BE SYSTEMATICALLY DIMINISHED

This means that we cannot harvest or manipulate ecosystems in such a way that productive capacity, ecosystem services, and diversity systematically diminish. While the first two system conditions refer to generation of substances into the ecosphere, this condition pertains to the destruction of the natural resources in such a way or to an extent that they cannot regenerate and be useful. An example is the overharvesting of forests to meet the current demand for wood, or for its conversion to paper. This also has a detrimental effect in reducing the sequestration of carbon dioxide. Human activities need to work in harmony with the cyclic principle of nature. Carbon dioxide emitted as a result of the combustion of fossil fuels is absorbed by vegetation and returned to the Earth's ecosphere. Overharvesting to meet the current demand may make it impossible for future generations to meet their requirements.

Other examples include the destruction of forests to convert the land for agricultural use. The increase in demand for food has forced some nations to trade off forests for agricultural land. This condition can also be considered in violation for the production of ethanol from corn. The land needed for this industrial crop could also be used for food crops. Overharvesting of fish may eventually deplete the fish population to a level below which regeneration to its original level may be impossible.

SYSTEM CONDITION 4—THERE MUST BE FAIR AND EFFICIENT USE OF RESOURCES WITH RESPECT TO MEETING HUMAN NEEDS

Basic human needs must be met with the most resource-efficient methods possible. Unless basic human needs are met worldwide through fair and efficient use of resources, it will be difficult to meet System Conditions 1, 2, and

3 on a global scale. As the human population continues to grow, it becomes more and more critical for the efficient use of the natural resources.

In late 2011, the human population exceeded seven billion, and this will require even more consumption of the Earth's natural recourses. This increase in population will lead to an increase in energy consumption, and with a finite quantity of available fossil fuels, more efficient use of coal, oil, and natural gas will be necessary. More efficient use also means an increase in renewable energy development and consumption. More efficient use of these energy resources will satisfy System Condition 4. At the same time, by adopting as much renewable energy as possible, less carbon dioxide will be emitted into the ecosphere and this will satisfy System Condition 1.

The use of plastic beverage containers is continuing to increase as the population grows and as more people drink bottled water. Unfortunately, in 2010, over 200 billion beverage containers were sold and over 130 billion of those containers ended up in landfills or were incinerated. Beverage container recycling rates have been declining nationally from 54% in 1992 to less than 33% today [6]. It is critical that more efficient use of plastic beverage containers is adopted in order to use less of the natural resources in their manufacturing and satisfy System Condition 4. In addition, there will be less plastic containers in the ecosphere and thus satisfy System Condition 2.

As the population continues to grow, another negative impact is the continued increase in fish consumption. Fishing is central to the livelihood and food security of 200 million people, especially in the developing world, while one of five people on this planet depends on fish as the primary source of protein. According to a Food and Agriculture Organization (FAO) estimate, over 70% of the world's fish species are either fully exploited or depleted. The dramatic increase of destructive fishing techniques worldwide destroys marine mammals and entire ecosystems [7]. More efficient harvesting of the fish population is critical in order to meet the needs of future generations and thus satisfy System Condition 4. By decreasing the fish population, the ability to increase the fish population will also diminish, thus violating System Condition 3. This is another example of showing that meeting System Condition 4 is critical to meeting the other three conditions.

SCIENTIFIC RATIONALE FOR THE NATURAL STEP

There are four basic scientific principles that form the foundation for the Natural Step Framework [1]:

1. Matter and energy cannot be created or destroyed (according to the first law of thermodynamics and the principle of matter conservation). This

means that the overall mass of the Earth remains constant. We have the same volume of matter now as we did 4.5 billion years ago. Therefore, matter—or the Earth's resources—only changes its form.

2. Matter and energy tend to disperse (according to the second law of thermodynamics). This means that sooner or later matter that is introduced into society will be released into natural systems. This is the underlying mechanism behind our experience that energy and material transformation operate to reduce the available energy in the system and increase the dissipation of matter throughout the system.

3. Material quality can be characterized by the concentration and structure of matter. What we consume are the qualities of matter and energy—the concentration, purity and structure of matter, and the ability of energy to perform work. Because nothing disappears and everything tends to disperse, a carpet turns to dust and a car turns to rust, and not the reverse. Dust does not reassemble into a carpet or rust into a car. As matter disperses, it loses its concentration, purity, and structure.

4. The net increase in material quality on Earth is produced by sun-driven processes. Photosynthesis is the only large-scale producer of material quality. While the Earth is a closed system with regard to matter, it is an open system with respect to energy. This is the reason why the system hasn't already run down with all of its resources being converted to waste. The Earth receives light from the sun and emits heat into space.

In addition to these scientific principles, the Natural Step conditions are influenced by the cyclic principle that can be summarized as follows: (i) waste must not systematically accumulate in nature, and (ii) the reconstitution of material quality must be at least as large as its dissipation.

THE NATURAL STEP RECENT PROJECTS [8]

Interface Reduces Water Use by 80% per Unit Since 1996

Interface, Inc. today uses only 20% as much water to make products as they did in 1996. This is an impressive accomplishment, especially for a materials manufacturer, and shows the benefits of looking at a business through the lens of sustainability.

Sixteen years ago, Interface, Inc. started on a journey led by their founder, R. Anderson. The company adopted a bold new vision—"To be the first company that, by its deeds, shows the entire industrial world what sustainability is in all its dimensions: People, process, product, place and profits—and in doing so—become restorative through the power of influence." Reaching this vision meant that Interface would need to evolve into an entirely different kind of organization, but there was no blueprint for this kind of organization in business.

Interface started by asking how they could translate the operations of nature into a model for business. They relied heavily on the Natural Step Framework to guide their thinking as they mapped out how to change their business. Interface ultimately developed the Seven Fronts of sustainability—seven key areas where Interface would focus to remake their company. Over the last 16 years, Interface has followed these Fronts and made progress, reducing the impacts of the company and its reliance on natural resources. This has included a strong focus on the manufacturing operations where they traditionally used large amounts of raw materials, energy, and water.

As Interface approached the management of their manufacturing operations through a sustainability lens, they were able to identify numerous opportunities to not only conserve resources but also cut costs.

Some savings have been the result of making fundamental shifts in how Interface manufactures its flooring products. For instance, engineers at Interface's Bentley Prince Street factory in California reconsidered the way they made carpet, which resulted in a reduction of the water used to manufacture each unit by 47%. The company simply moved away from the energy-, water-, and chemical-intensive dye-injection and yarn-dye methods for adding colors and patterns to their products. Dye injection involved using a large machine, much like a big ink-jet printer, to apply colors to long lengths of carpet. The yarn-dye method involved soaking fibers in dye solution before tufting it into carpet. In pursuit of their sustainability objectives, Interface Bentley Prince Street totally phased out the use of dye-injection carpet printing in 1999 and reduced carpet made with the yarn-dye process from 45% of its output in 2001 to 1% in 2009.

As a substitute, the company shifted its focus to two alternative processes—piece dyeing and solution-dyed yarns. The piece dyeing process involves manufacturing a blank slate of white carpet and then adding patterns and colors on a made-to-order basis. This process of customization is far more energy and resource efficient than the sweeping dye-injection method.

The company also reconsidered the process of creating the carpet yarn itself. Instead of soaking fibers in dyes, the switch was made to the use of yarn filaments that are extruded from solutions already impregnated with pigment. This dyeing method drastically reduces the amount of water, energy, and chemicals required in the process, and at InterfaceFLOR's modular carpet factory in Georgia, this change in the manufacturing process resulted in a savings of over 88% in water per unit of production.

Interface has also taken advantage of the low-hanging fruit available to them by making easy changes in building operations. At InterfaceFLOR in Thailand, they have reduced total water usage by 30% through the installation of water-efficient toilets, showers, and sinks. Outside, they selected plants for their landscape plan that only required one year of temporary irrigation.

Interface's significant reduction in water use is just one example of how managing the environmental impacts of business can bring about meaningful change. By relentlessly pursuing sustainability-focused innovation, Interface is racing to fulfill its long-term vision of becoming a "restorative company," and is continuing to set the pace as a world leader in next-generation manufacturing.

VinylPlus: The European PVC Industry's Voluntary Commitment to Sustainable Development

On June 22, 2011, the European polyvinyl chloride (PVC) industry launched VinylPlus—a new voluntary commitment to enhance the sustainable production and use of PVC by 2020. It's no secret that PVC is a controversial material that has both critics and supporters, so when an entire industry consortium mobilizes around a voluntary commitment such as this, it is surely a positive sign that sustainable development is being taken seriously in the board rooms of major corporations.

On top of this, the VinylPlus initiative represents an expanded scope and level of ambition over prior efforts. Targets are derived from the Natural Step System Conditions for sustainability and build on previous work in the sector (see links in the succeeding text).

Over the last two years, through a series of activities, the Natural Step has worked with the Board of the initiative to review and map out the challenges for PVC, offer capacity-building workshops, gather stakeholder views, and provide recommendations on the formulation of the new industry charter.

They used the ABCD planning process to help industry leaders apply the framework and arrive at the priorities. They did this by first beginning with a vision of where the industry wants to head, identifying challenges that must be overcome, generating ideas on what needs to be done, and finally arriving at priorities and targets for the period to 2020.

Before deciding on the priorities and actions to be undertaken, they contacted 113 stakeholders across Europe on behalf of the Board of Vinyl 2010 with a briefing on the challenges they had identified together with industry representatives. These stakeholders included representatives of the public, PVC users, regulatory and policymaking institutions across the EU, and sustainability-related NGOs.

Following feedback from stakeholders, the priorities and concerns were reported to the Board of Vinyl2010 and used in the formulation of exact goals and targets.

They were pleased to see an increased level of ambition in VinylPlus and believe it sets out a clear roadmap for the industry. Proof will nevertheless be in the action, and it will be judged by whether targets are met and stakeholders feel their views are being heard.

Moving forward, the Natural Step's role will be to follow the new initiative in the following capacity:

1. *Critical friend*—challenging and advising the industry to make progress while ensuring the initiative retains the direct link to the Natural Step System Conditions for sustainability
2. *Stakeholder intermediary*—encouraging constant external monitoring and communication with external stakeholders
3. *Capacity builder*—supporting the industry to integrate sustainability principles into their operations in order to achieve the goals set out in the initiative

The VinylPlus Board welcomes feedback on the initiative and particularly our role as outlined in the preceding text.

Real Change for the Rhine

At 1320 km in length (820 miles), the Rhine is a modest river by global standards. But from an economic standpoint, it is one of the most important in the world, as it serves as a gateway to five of the most prosperous countries in Northern Europe. Each year, ships log 185,000 trips traveling to and from Rotterdam (the Netherlands) and Duisburg (Germany), the largest seaport and inland ports of Europe, respectively.

The Province of Gelderland in the Netherlands sits at the river's delta at the North Sea, and its government has become concerned about the long-term sustainability of the region in the wake of widespread flooding and disruptions in the area. After years of trying short-term technical interventions to "fix" this problem, there is now an awareness that traditional approaches are out of date and that land use, river management, and flood protection strategies should focus more on long-term interventions. This has triggered a series of international debates, developments of innovative techniques, participation and decision-making processes, as well as an engagement with the Real Change Programme, a research initiative co-lead by the Natural Step.

Through Real Change, the Province has partnered with the Department for Sustainable Management of Resources at Radboud University Nijmegen to facilitate action research focusing on sustainable development of the Rhine delta region. The team at Radboud University Nijmegen is introducing TNS concepts and methodologies to the civil servants of 15 communities along the Rhine–Waal river sections and leading visioning sessions with "Community of Practice" groups. They will be incorporating this learning into 48 land-use projects by building upon existing flood protection strategies and then studying how to best apply alternative land use, ecological rehabilitation, and reduction of greenhouse gas emissions (GHGs) techniques to further the development of a sustainable regional economy.

River deltas are some of the most densely populated areas in the world. The abundance of fertile grounds, wildlife and freshwater, and sea trade led to their early settlement, and the Rhine delta is no exception. It is still a densely populated area of culture and art, and the competing interests of people, planet, and profit makes the challenge of implementing sustainable development practices even more challenging, and we look forward to sharing our successes and challenges with the Natural Step network via Stepping Stones in the near future.

Dow Measures Up

By any measure, Dow is a sustainability leader.

Dow has been listed in the Dow Jones Sustainability Index nine times. They have won "green" awards from Michigan to China for their wide array of products from insecticides to solar panels. And in 2006, CEO Andrew Liveris enthusiastically announced seven new sustainability goals for 2015 even *after* the company had already met or exceeded many of its 1996 sustainability goals. In that initial 10-year span, they saved more than five billion US dollars in bottom-line costs.

Yet probably, their most impressive feat to date might be their array of clear and simple bar graphs that dot the end of each of their quarterly (yes, quarterly!) sustainability reports. These simple beige and green charts aren't flashy—they probably were produced right in an Excel spreadsheet. But it is the function of these graphs that make them so meaningful and compelling, based on what they represent.

First, whether one particular graph is showing their product safety leadership or the intensity of GHGs, you can immediately see the upward or downward trajectory the company is on. Their goals are always marked with a simple black line, and for the most part, the green and orange bars show that they are hitting their intended targets. *Dow's progress is clear.*

Second, Dow may have some of the most complete, public, and long-range sustainability data available. It is immediately evident that the company keeps great records of the data that they produce. This meticulous study certainly seems central to the corporate culture, with good reason—Dow is driven by the exacting art and science of chemistry, which demands precision from heaps of data. *Dow's progress is credible.*

The data are so complete that their sustainability managers are able to make sense of specific anomalies when they occur. For instance, in 2009, the company's intensity of GHGs (the amount of energy used to create a single pound of finished product) increased by 5%, which went in the opposite direction after four years of downward progress. They were able to attribute this slide to lower operating efficiencies of their production facilities during the worldwide recession. In short, they could see exactly what the effect underutilizing their facilities had on their overall emission levels—a fantastic feat for any company as large as Dow. *Dow's progress can be meticulously tracked.*

There isn't a company in the world that can claim to be operating in a "sustainable" way, and Dow most certainly has a long way to go to get beyond doing "less bad" and claim that they have no negative impacts on social and ecological systems. But one thing is clear: with the help of their simple bar graphs, Dow will probably know it the exact moment when they do get to that point!

Low Fat Carbon in the South of New Zealand

Serra Natural Foods is a New Zealand-based manufacturer of organic dairy products with an established presence across New Zealand and Australia and a growing market share in Canada and the United States. The company has found great value in using the Natural Step Framework to guide its brand image and business strategy (http://www.naturalstep.org/en/new-zealand/low-fat-carbon-south-new-zealand).

> Consumers today are very sophisticated and discern subtle messages about brands in all manner of ways" says CEO Jim Small. Because health and sustainability are important aspects of the brand personality, it is essential that our company acts in a way that is consistent with that brand image. Even when times are tough and there are less resources to invest in new programs, The Natural Step awareness persists in keeping the business facing in the right direction.

Serra has recently reengineered its domestic sales and distribution system in New Zealand in order to make it more efficient and effective. Internationally, the focus of its export sales has turned to frozen yogurt, which is easier to transport via lower sea freight than fresh yogurt. The company now also includes its carbon emissions data on its profit and loss statements so that it is directly linked to performance measurements.

Nike's Core Values

Nike credits the Natural Step, Business for Social Responsibility (BSR), Deloitte & Touche, and the World Economic Forum as collaborators that helped Nike define its strategic path towards sustainability, and they are very proud to have played such a larger part in Nike's ongoing efforts.

This step is a big one and it reflects a cultural shift coming from the core values of the organization. CEO Mark Parker explained:

And for all the athletic and cultural and financial successes of the company, I believe our work in sustainable business and innovation has equal potential to shape our legacy. For that to happen, we have to focus on the lessons we've learned:

- *Transparency is an asset, not a risk.*
- *Collaboration enables systemic change.*
- *Every challenge and risk is an opportunity.*
- *Design allows you to prototype the future, rather than retrofit the past.*
- *To make real change, you have to be a catalyst.*

As part of this work, Nike tested the real business impact of the changing world on NIKE, Inc., and explored how the consumer brand could thrive in a sustainable economy. They concluded that it needs to continue to refine its existing business model while simultaneously looking at new ways of doing business.

As Nike entered 2010, the work continued, with a number of ambitious goals:

1. Put investing in sustainability as a key innovation/R+D priority on consumer brands' agendas.
2. Fast-track innovation through investment and collaboration.
3. Launch the GreenXchange as a platform for enabling the sharing of intellectual property to fast-track changes efficiently.
4. Build an advocacy agenda to push for large-scale policies and investments in sustainable innovation.

Nike's commitment did not come about overnight. They had been transforming how the organization saw its place in the market and the world for 10 years. They are proud to be part of that ongoing transformation and hope other organizations would recognize and rise to the challenges they face with similar optimism and enthusiasm.

Pratt & Whitney Canada's Sustainability Journey Aerospace manufacturer Pratt & Whitney Canada (P&WC) wanted to explore how building on existing sustainability initiatives, towards a proactive approach to sustainability, could help capture additional market share and drive innovation for the company.

In early 2011, P&WC engaged with the Natural Step Canada to undertake an organizational review, create a vision and goals for the organization, and develop a sustainability roadmap to guide their progress toward minimizing their environmental impact while capturing value for the organization.

To begin, the Natural Step Canada undertook an organizational review of P&WC's operations and culture. This helped determine the business's ability to capture the sustainability value proposition and identify opportunities and threats that might exist. Information was collected over several months through document review, interviews, and surveys and was then analyzed to determine internal attitudes towards sustainability and how staff felt about the organization taking a sustainability-driven leadership role within the aerospace industry. In addition, an external scan conducted by the Natural Step Canada also provided P&WC with a benchmark report about the aerospace industry's sustainability opportunities.

Next, during a planning workshop, a group of executives, senior managers, and members of the internal sustainability team at P&WC was introduced to the Natural Step Framework. Armed with a common language and understanding of sustainability, the group worked to identify some of P&WC's key sustainability challenges as well as strategic goals to address them. This process enabled the development of a series of transition strategies and key milestones to guide P&WC's sustainability efforts over the next few years.

A few months later, in alignment with the roadmap from the workshop, P&WC had a more defined approach, sustainability goals, and strategy teams in place. One of the organization's priorities became to embed all dimensions of sustainability into its corporate culture while also working towards more sustainable products and operations.

The feedback from the company resulted in the following: "At Pratt & Whitney Canada, we are firmly committed to ensuring that our products are designed, produced, and operated to enable minimization of environmental impacts throughout their entire life cycle. The Natural Step helped us define a clear sustainability strategy and key priorities as we continue to strive to become 'the best aerospace company FOR the world'."

REFERENCES

1. Nattrass B, Mary A. *The Natural Step for Business: Wealth, Ecology, and the Evolutionary Corporation*. Gabriola Island, BC: New Society Publishers; 1999.

2. Available at http://www.thenaturalstep.org/en/our-story. Accessed 2013 Jun 29.

3. Available at http://strategiesforsustainability.blogspot.com/2006/04/funnel.html. Accessed 2013 Jun 29.

4. Harris R. Global warming is irreversible, study says. *All Things Considered*, NPR, January 26, 2009.

5. Meadow D, Randers J Meadows D. *The Limits to Growth: The 30-Year Update*. White River Junction, VT: Chelsea Green Publishing Company; 2004.

6. Available at http://www.asyousow.org/sustainability/beverage.shtml. Accessed 2013 Jun 29.

7. Available at http://www.un.org/events/tenstories/06/story.asp?storyID=800. Accessed 2013 Jun 29.

8. Available at http://www.naturalstep.org/en/current_projects_all. Accessed 2013 Jun 29.

CHAPTER 5

Eco-effective Versus Eco-efficient: Sustainability Versus Being "Less Bad"

In 1973, the Organization of the Petroleum Exporting Countries proclaimed and oil embargo in response to the US decision to assist Israel in its war with Egypt. This embargo affected not only the United States but also its Western European allies. Although the United States was energy self-sufficient through 1950, it was now importing about 35% of its oil. While this embargo was lifted the following year, the US administration was still concerned that something similar could happen again.

In 1975, the US Congress passed the Corporate Average Fuel Economy (CAFÉ) standards to double the average then fuel economy of 13.5 miles per gallon to 27.5 miles per gallon by 1985. These standards started at 20.0 miles per gallon in 1980 and increased by 2.0 miles per gallon each year until they reached 27.5 miles per gallon in 1985. These standards remained the same for the next 25 years, after which they will increase in order to make the auto industry more efficient.

Making automobiles more efficient with respect to gasoline consumption is the way to go. It is critical that less oil be consumed as it is a nonrenewable energy source with depleting inventory worldwide. To become more eco-efficient, we can increase the automobile fuel economy by, say, 50%, which will reduce the consumption of oil. But with growing economies in China and India, what if the number of automobiles in the world doubles? The total consumption of oil will continue to increase.

About the same time, in 1987, the UN World Commission on Environment and Development summarized its meeting in a report titled *Our Common Future* and emphasized that industrial operations should be encouraged to reduce the consumption of nonrenewable natural resources, generate less waste and pollution, and minimize irreversible negative impacts on human

Practical Sustainability Strategies: How to Gain a Competitive Advantage, First Edition. Nikos Avlonas and George P. Nassos. © 2014 John Wiley & Sons, Inc. Published 2014 by John Wiley & Sons, Inc.

health and the environment. It was five years later that the Business Council for Sustainable Development (BCSD), a group of 48 industrial companies, officially adopted the word "eco-efficient." At the 1992 Earth Summit in Rio de Janeiro, the BCSD brought this concept of eco-efficiency to the representatives of 167 countries. Industrial machines would now become cleaner, faster, quieter, and thus more efficient.

Another version of this same concept is *lean manufacturing*, which focuses on more efficient operations while generating less waste. Lean manufacturing is actually based on the Toyota Production System developed by Toyota Motor Corporation [1]. Eco-efficiency also led to the initiative of three Rs, *reduce*, *reuse*, and *recycle*. Since then, this initiative has grown to five Rs with the addition of *redesign* and, to a lesser extent, *regulate*.

While eco-efficiency is extremely important in leading to a more healthy and sustainable environment, it is a partial solution. Eco-efficiency really means being *less bad*. We must strive for a total solution and be *eco-effective*. Reducing toxic emissions like dioxins from a power plant by adding new pollution control equipment is eco-efficient. However, preventing any toxic emissions is eco-effective. Recycling of plastic beverage bottles is eco-efficient, but banning plastic beverage bottles and replacing them with reusable bottles is eco-effective. Let's consider a few examples of how eco-efficiency can be improved by eco-effectiveness.

FUEL EFFICIENCY

Going back to the beginning of this chapter, there is a major effort to increase the fuel economy of automobiles worldwide. In the United States there are new regulations that will require the CAFÉ standards to increase to 36.5 miles per gallon by 2016 and 54.5 miles per gallon by 2025. If the number of cars on the road in 2025 in the United States is the same as 2011, the consumption of oil will be about 50% in 2025 as compared to 2011. However, let's look at the rest of the world. The number of cars in the world increased by 3.5% to over one billion in 2011 and is expected to exceed 2.5 billion by 2050 [2]. Half of this growth was from China, which increased the number of cars on the road by 27.5%. Most of the growth in cars on the road by 2050 will be from China and India.

The fuel economy standards in China were already at 36.5 in 2009 and are set for 42.2 by 2015 [3]. In India, the fuel economy standards will increase from the current average of 32.2 miles per gallon to 40.7 miles per gallon by 2015 [4]. If we can assume that the overall fuel economy will double by 2050, the consumption of oil will still increase by 25%. Despite being eco-efficient, more oil will be consumed and more carbon dioxide will be emitted. We need to be eco-effective.

The route to eco-effectiveness is to eliminate fossil fuels altogether. One way of accomplishing this task is to switch to manufacturing electric automobiles and eventually remove all the gasoline-powered cars off the roads. Battery-operated cars, like gasoline cars, have a limited range that today is much less than the gasoline cars. Consequently, the batteries must be charged at shorter intervals, but the charging system is critical to eco-effectiveness. If the power required to charge the batteries is generated from a coal-fired or natural gas plant, fossil fuels are being consumed and carbon dioxide is being emitted. So even if less fuel is consumed as compared to an eco-efficient, gasoline-driven vehicle, this is still being eco-efficient.

An electric car charging network like Better Place [5] possesses several features that make this system eco-effective. This system consists of a network of charging stations geographically located so that the battery range will allow the electric car to travel between stations. Once the car arrives at the station, the batteries are changed out for fully charged batteries in a matter of minutes, and the driver continues the trip. The key to making this system eco-effective is that the batteries are charged by employing renewable energy like solar or wind. The first system deployed by Better Place is in Israel, where the batteries are charged with solar energy [6]. In a small country like Israel or Denmark, it is possible to develop an infrastructure of charging stations that makes it possible for a driver to travel throughout the country and find conveniently located charging stations. This is truly being eco-effective.

COMPUTING EFFICIENCY

Since the beginning of transistor-powered computers, the major manufacturer of the transistors has been Intel, a company founded by Gordon Moore. The business model of this company and its competitors was to double the number of transistors in a computer every 24 months, and along with making them faster, the computer's processing capacity would double. This was eventually known as Moore's Law [7], although an Intel executive modified the doubling period to 18 months and this number is quoted more often.

Based on Moore's Law, every 18 or 20 months, the computers become more efficient and thus more can be accomplished with less. This is another example of eco-efficiency. Thus, businesses that require the most efficient computers, like financial services companies, will replace their computers every 18–20 months, primarily to be more efficient. But what happens to the older, slower computers? Many of them are transferred to another organization that doesn't require the latest computer efficiency, and in turn theirs are sent to less proficient users. For example, some universities always provide the most efficient computers for the student computer labs or IT classes. The replaced computers would then be given to the faculty and staff, and their old ones would be sold for home use or to local schools. In any event, while eco-efficiency is being exercised, more and more computers are being placed in use.

In order to be eco-effective, a different business model must be introduced so that additional computers are not introduced into society if not needed. An alternative would be to manufacture computers in which the processor can be replaced with a newer, faster version. There is no need to replace the keyboard, the monitor, and other components if no improvement in these components is anticipated. This model would require less resources like metal and plastic and would create less waste. This model would be eco-effective.

MORE DURABLE BRAKE PADS

Brake pads for automobiles were originally made from materials like carbon black, graphite, and asbestos. As the brake pads are employed for their function, particles of the composite material are emitted into the atmosphere and the landscape. In the 1980s, the brake pads were made smaller, which required the composition to be more durable. So the pads were made from carbon black, bronze, and resins. More recently, the brake pad composition has changed to fiberglass, Kevlar, ceramic, and some asbestos [8]. In each of these types of brake pads, the material of composition is eventually dispersed in the atmosphere. The focus has been to be more eco-efficient and manufacture more durable brake pads so that less material is used in the manufacturing process and less toxic material is emitted into the atmosphere and landscape. The US government is even trying to legislate regulations for the prevention of the emission of toxic substances like asbestos [9].

In order to be eco-effective, the brake pads should be manufactured using materials that are not only nontoxic but also renewable and/or compostable. The disintegrated brake pad material is found primarily in the landscape along roads. With a durable and compostable brake pad material, the process would become eco-effective.

INCREASE POLYMER RECYCLING

The amount of plastic goods being manufactured is continuing to increase, mostly made from oil. The continued increase in production is due to increased demand and the deficiencies in adequate recycling. What recycling of plastic is accomplished, some people call it "down-cycling" because in most cases the recycled plastic is then used for some lesser application. Some uses for plastic products have specifications that require the product to be manufactured from petroleum rather than recycled material.

In order to be more eco-efficient, efforts are being made to recycle more plastic so that less plastic is manufactured from petroleum. In some cases, the product can be made from a combination of virgin plastic and recycled plastic. Some beverage bottles are currently manufactured from that combination with the recycled content as high as 30%. This is going in the right direction, but it is not enough. It is being "less bad."

In order to be eco-effective, we need to think about fulfilling the demand for the plastic and not manufacturing any additional plastic. Rather than "down-cycling," we need to find ways of reusing the plastic material. Some products manufactured from plastics might be disassembled at the end of their useful life and reassembled for some other application. Another possibility is to develop a process that allows for continuous recycling of the same plastic material. Approaching that concept is PepsiCo, which is constantly looking for ways to be eco-effective. In 2011, PepsiCo introduced a plastic bottle made from 100% polyethylene terephthalate [10]. How many times that bottle could be recycled has not yet been determined.

Another way to become eco-effective is to manufacture the bottle from renewable resources rather than nonrenewable. Both PepsiCo and Coca-Cola are developing a beverage bottle made entirely from plants. In 2009, Coca-Cola announced that it is manufacturing a beverage bottle made from 30% plant-based plastic. The two companies are racing to see which will be first with a 100% plant-based plastic bottle [11]. This is truly being eco-effective.

REDUCED SEWAGE EFFLUENT

One of the industries that has been of concern for some time relative to waste effluent is that of textile dyeing and finishing. The whole cycle of finishing consists of mechanical and chemical processes, which are used depending on the kinds and end uses of the fabric. Mechanical processes include drying, calendaring, schreinerizing, embossing, and sueding, and chemical processes include the application of special substances on the fabric and impregnation with size, starch, dextrin, and other polymeric substances [12]. The effluent can include chemicals like halogenated organic compounds, heavy metals (chromium, copper, zinc), surfactants, and salts.

In order to be eco-efficient, it is critical to minimize the volume of the effluent and/or reduce the concentration of the pollutants in the effluent.

When assessing options to minimize wastewater streams, one environmental agency [13] suggests the following methods:

- Minimize machine cleaning through better maintenance and production planning.
- Optimize production to reduce liquor ratio.
- Optimize and reduce the number of rinses.
- Optimize cycle times and job turnaround.
- Use lower liquor ratio machinery.
- Reduce reprocessing through better quality controls.
- Combine rinses with scours.
- Scours may be done in dyebath.
- Recycle steam condensate back to boilers.
- Recycle cooling waste to use as hot/mixed hot–cold fill.
- Recycle rinses as feeds for dyebaths and scour baths.
- Recycle "clean rinses."

These recommendations should be implemented in order to be eco-efficient, but this, again, means being "less bad."

What should be done to make the textile dyeing and finishing industry eco-effective? Industry engineers and chemists should find additives, such as vegetable dyes, that are compostable so that the residue from the wastewater treatment system can be returned to the earth. Employing compostable dyeing and finishing chemicals may also make the final product compostable at the end of its useful life if the fabric is a natural material like wool or cotton.

MORE EFFICIENT CLOCKS

Here is another example that was first presented by Denis Hayes, President and CEO of the Bullitt Foundation. We have been using electric analog clocks for many years, and they are not very efficient. A typical electric clock requires about four watts of power that translates to 96 Watt-hours per day. In one year, the clock will consume about 35 kilowatt hours of energy. Assuming there is one electric analog clock per person in the United States, a total of about 1000 megawatts of power would be required, and this much power consumes about 100 railroad cars of coal per day. How can we make these clocks more efficient?

Electric wristwatches operate with small batteries that possess minimal power. The same number of these watches would require about three kilowatts in total. This amount of power from a coal-fired plant would need one railroad car of coal every five years. This is an excellent example of eco-efficiency, but how do we make these clocks eco-effective?

In a two-step process, the operation of electric clocks can be made eco-effective. The first step would be to make the electric clock as energy efficient as a battery-operated wristwatch. The second step would be to connect the clock to a small solar/battery system so that all the power required for the clock would be provided by the solar cell. However, a battery charged by the solar cell would be needed to operate the clock during the evening. Now this system is eco-effective.

CRADLE TO CRADLE

Most systems designed for eco-efficiency are based on getting more output from less input. Some of these systems may be based on dematerialization—or servicizing (see Chapter 6), extended life of the product, multiple recycling, reducing pollution or toxicity, or obtaining more output from the resources. These strategies are all based on a linear flow of materials from their source to the landfill, or from cradle to grave. McDonough and Braungart [14] proposed a five-step system that leads to eco-effectiveness, or cradle to cradle.

Step 1: Get Free of Known Undesirable Substances   There are many substances that are known to be harmful and thus undesirable in products utilized by businesses or consumers. Most companies are not aware of the toxicological impact of substances in products through their supply chain. However, they should have a general knowledge of the most dangerous substances in their products. These are referred to as X-substances when referred to eco-effectiveness. For companies like this, a first step in moving toward eco-effectiveness is to find replacements for the X-substances in their products. This includes substances like mercury, cadmium, lead, and polyvinyl chloride (PVC) that are known for suspected carcinogens, mutagens, or endocrine disruptors. Removal of X-substances is almost always a step in the right direction, but such an approach has to be applied carefully to ensure that replacement substances are indeed better than those that are replaced.

Step 2: Follow Informed Personal Preferences   Once the most undesirable substances have been removed from a product, the next step is to begin to make educated choices about those substances that should be included in the product. The best way to do this may be to have a detailed knowledge about the impacts of a particular substance on ecological and human systems throughout its life cycle. But this is often impractical or even impossible. Furthermore, different substances may have different types of impacts on different end products. Some products may require an additive that could be one of several different materials. Each of these may have a different impact such as persistency in the environment, contribute to climate change, or harm plant life. Without a detailed scientific knowledge of a substance's toxicological profile and its fate throughout the life cycle of a product, these decisions can be difficult to make. Sometimes it may be necessary to bring a product to the market without really knowing the eventual impact. Without complete knowledge of these substances, the best way to make decisions about which chemicals and materials to include in a product comes down to personal preferences based upon the best available information. Though decisions guided by personal preferences may not always result in the most eco-effective design choices, they generally will result in a product that is at least less bad than its predecessors. Although it is "less bad," it is on its way to become eco-effective.

Step 3: Creating a "Passive–Positive" List   This step includes a systematic assessment of each ingredient in a product to classify it according to its toxicological and ecotoxicological characteristics, especially its capability to flow within biological and technical metabolisms. For products of consumption, the criteria to examine should include impacts like toxicity to humans (acute, delayed, developmental, reproductive), aquatic toxicity, persistence and bioaccumulation in nature, sensitization potential, mutagenicity, carcinogenicity, and endocrine disruption potential. Based upon the assessment of a material or chemical according to these criteria, a passive–positive list can be generated that classifies each substance according to its suitability for the biological metabolism. This list can be used to determine the degree of additional optimization necessary for a particular product to be a true product of consumption. One can also use this process for service-type products, although the impacts may be different. Cadmium, for instance, is a highly toxic heavy metal and is often applied in photovoltaics in the form of cadmium telluride that may not necessarily be bad. Although cadmium telluride is not an ideal substance from an ecological perspective, its application in photovoltaics as a product of service may be considered acceptable until something better is found and employed.

Step 4: Activate the Positive List   In this step, the objective is to determine which component materials provide a positive technical or biological nutrient. In Step 3, the objective is to optimize the amount of the material to minimize its impact, whereas in Step 4 the objective is to optimize the component to its fullest extent. An example is an upholstery fabric whose constituent materials are positively defined as biological nutrients. This was designed by McDonough and Braungart and others for a German company and is a completely biodegradable and compostable fabric. Each component was selected according to the manufacturer's positive listing methodology for its positive environmental and human health characteristics and its suitability as a biological nutrient.

Step 5: Reinvention The concept of reinvention addresses the interconnected nature of ecological, social, and economic systems by considering the biological and technical impacts of existing product and service forms. Strategies for reinvention view products from the perspective of the services they provide and the needs they fulfill for customers. They also consider the broader context of social and ecological systems. While most companies try to design a product with minimal negative impacts, reinvention considers the design of a product with maximum positive impacts. An example may be to design a car whose emissions can be considered nutritious for nature or industries. Perhaps it could burn a fuel that emits water vapor that can be turned back to water and used for some end purpose. Or perhaps the carbon dioxide released from the exhaust can be captured and converted to carbon, which can be used in the manufacture of rubber tires. As more cars are manufactured and driven in the world, more nutrition will become available.

This five-step process to develop an eco-effective product, system, or service will take a trial and error process along with time, effort, money, and creativity.

DON'T TAKE IT TO THE EXTREME

If one gives considerable thought to this sustainable strategy, it should be possible to design a system that is extremely eco-effective, but not very desirable. One might design a pill that provides all the nutrients and vitamins needed for sustaining a healthy life. You could call it a dinner pill. But should we go to that extent and eliminate the pleasure of eating?

REFERENCES

1. Available at http://www.toyota-global.com/company/vision_philosophy/toyota_production_system/. Accessed 2013 Jun 29.
2. Available at http://www.huffingtonpost.ca/2011/08/23/car-population_n_934291.html. Accessed 2013 Jun 29.
3. Available at http://www.edmunds.com/autoobserver-archive/2009/05/china-sets-422-mpg-fuel-efficiency-standard-as-it-seeks-to-curb-oil-consumption.html. Accessed 2013 Jun 29.

4. Available at http://www.greencarcongress.com/2011/05/india-20110504.html. Accessed 2013 Jun 29.

5. Available at http://www.betterplace.com/. Accessed 2013 Jun 29.

6. Available at http://www.betterplace.com/global-progress-israel. Accessed 2013 Jun 29.

7. Available at http://www.intel.com/content/www/us/en/history/museum-gordon-moore-law.html. Accessed 2013 Jun 29.

8. Available at http://www.ehow.com/about_4611238_brake-pads.html. Accessed 2013 Jun 29.

9. Available at http://www.saferchemicals.org/resources/chemicals/asbestos.html. Accessed 2013 Jun 29.

10. Available at http://www.pepsico.com/PressRelease/PepsiCo-Beverages-Canada-Unveils-the-7UP-EcoGreen-Bottle-Canadas-First-Soft-Drin07132011.html. Accessed 2013 Jun 29.

11. Available at http://www.nytimes.com/2011/12/16/business/energy-environment/coca-cola-and-pepsico-race-for-greener-bottles.html. Accessed 2013 Jun 29.

12. Available at http://www.textileschool.com/School/TextileFinishing/FinishingProcess.aspx. Accessed 2013 Jun 29.

13. Environment Protection Authority – State Government of Victoria. *Environmental Guidelines for the Textile Dyeing and Finishing Industry*. Melbourne: Environment Protection Authority – State Government of Victoria; June 1998.

14. McDonough W, Braungart M. *Cradle to Cradle*. North Point Press, New York; 2002.

CHAPTER 6

Extended Product Responsibility and "Servicing"

Over the past generation, the economies of the United States and other wealthy industrialized states have undergone significant structural changes. Services have attained new prominence, and the relative contribution of traditional manufacturing to these economies has diminished. These changes have created enormous opportunities for entrepreneurs and new national wealth on the one hand and huge social costs attendant to the decline of traditional industries and challenges for public policy on the other.

Because the human impact on the environment is intimately linked to economic activity, these changes present both challenge and opportunity for environmental policy. The structural changes producing a service and information-led economy are often presented in environmental terms as gradually divorcing economic growth from material and energy throughput and environmental burden. The idea of a *functional* economy—in which the focus of consumption is not goods per se, but the services that those goods deliver—has been associated with the idea of eco-efficiency. In a functional economy, commercial and domestic consumers buy cleaning services instead of washing machines, document services rather than photocopiers, and mobility services rather than cars.

Systematic analysis of the environmental implications of a service and information-led economy is just beginning. *It is clear that the simplest and most optimistic view—a service economy is inherently a clean economy—is insufficient and incorrect.* Instead, the service economy is better characterized as a value-added layer resting upon a material-intensive, industrial economy. All else equal, economic growth in services may be less environmentally problematic than growth in manufacturing. But that is not sufficient when society already exceeds environmental limits in a number of crucial

Practical Sustainability Strategies: How to Gain a Competitive Advantage,
First Edition. Nikos Avlonas and George P. Nassos.
© 2014 John Wiley & Sons, Inc. Published 2014 by John Wiley & Sons, Inc.

ways. *If services are to produce a greener economy, it will be because they change the ways in which products are made, used, and disposed of—or because services, in some cases, supplant products altogether* [1].

The concept of services *supplanting* products has been considered for some time. Two examples are the "dematerialization" potential of information technology in reduced travel and transport of physical goods. However, not all products can be dematerialized. For example, the provision of basic needs such as food, shelter, and clothing requires the mobilization of an irreducible physical minimum of material and energy, but the information technology that may make dematerialization possible rests upon an extensive, sophisticated manufacturing and maintenance infrastructure. While most examples of dematerialization were pursued for economic reasons, more recent efforts are for environmental reasons.

When one talks about dematerialization and providing a service, it should be made clear that this type of service is different from the traditional services like banking, accounting, hair dressing, and reservations. Consequently, the term that was created to make this differentiation is "servicizing" and the process can be called "servitization." Rather than selling and buying a product, the property rights can be some diversification of leasing, pooling, sharing, or take-back.

Servicizing can also be a driving force for extended product responsibility (EPR). This is the principle whereby the process participants along the product chain or life cycle share responsibility for the life cycle environmental impacts of the whole product system, including upstream, production, and downstream impacts. This should result in lower life cycle environmental impacts for products or product systems. The link between servicizing and EPR is that both require manufacturers or service providers to extend their involvement with, and responsibility for, the product to phases of the life cycle outside the traditional seller/buyer relationship. If servicizing contains within it potential environmental benefits, it is because this altered relationship with the product drives superior environmental performance—in short, because servicizing drives EPR.

When companies manufacture a product for a particular application or function, it transfers ownership of the product to the customer who takes over all responsibilities. The manufacturer may provide a warranty for its operation over a particular period of time, but the ultimate responsibility for recycling, reuse, or disposal belongs to the customer. There may be, however, certain advantages of extending the product responsibility along the supply chain or the product's life cycle.

For many of the manufacturing businesses today, the business model is based on utilizing the natural resources and energy to manufacture a product that provides some useful function to the consumer. With the exception of some warranties, the manufacturer usually has no other obligations for that product. The consumer uses it throughout its useful life and then disposes of it, with the product eventually finding its way to the local landfill, even after some recycling.

Employing this servicizing paradigm, companies can sell the function of the product rather than the product itself and maintain ownership of the product throughout its useful life. Under such a scenario, the manufacturer would be more inclined not to consider planned obsolescence and manufacture a more durable product. This would result in fewer products manufactured, less resources employed, and less waste created.

SELL ILLUMINATION

A simple example of this business model is in the manufacturing and use of incandescent light bulbs. Manufacturers like Phillips and GE sell light bulbs, through a supply chain, to an end user, such as a school. But the school does not really want light bulbs; it wants the function of light bulbs—illumination. So why not sell illumination? This could be accomplished by providing illumination in terms of watts per hour or lumens per hour, or some other measure with the ultimate cost to the school being the same or less.

In the original business model, the sale of light bulbs is a revenue item, and it shows up on the company's profit and loss statement as a sales. In business terminology, it is considered a profit center for the company. By selling illumination, this service becomes the profit center, and the light bulb itself becomes a cost center, thus providing an incentive for the company to reduce that cost. But how? Currently, when a light bulb burns out, it is because the tiny metal filament in the bulb is broken. The owner of the light bulb throws it out and replaces it with a new one, even though the only thing wrong with it is the tiny filament inside. When the bulb is disposed, the metal, ceramic, and glass are all discarded even though it is only the filament that went bad.

What if GE or Phillips could manufacture a light bulb where the glass bulb itself could be safely removed and the broken filament replaced? Now the only resource necessary to produce a new light bulb is the tiny filament and the only waste that would be created is the broken filament. Actually, this tiny filament need not be waste if it is eventually melted and reformed into a new filament. In theory, once all the light fixtures are fitted with this new light bulb, there would be no need to ever manufacture another light bulb. Of course, this new reusable bulb would cost a little more to produce, but the ultimate use of resources and creation of waste would approach zero.

SELL A PAINTED CAR

There are some real examples of how this new business model can be implemented. The Ford Motor Company has an assembly plant in England where the cars are painted using product supplied by DuPont. DuPont's objective is to sell as much paint as possible to Ford, as selling paint is how it makes money, that is, its profit center.

A few years ago, Ford management approached the DuPont salespeople and asked them to sell Ford a painted car instead of the paint. Upon agreement with this suggestion, DuPont assumed the responsibility for painting the cars, and in providing this service, paint was no longer the profit center as it became a cost item. DuPont would now be paid on a per-car basis rather than on a per-gallon-of-paint basis. DuPont then worked on a new formulation to use less paint while meeting Ford's specifications and improved the efficiency of the spray guns in order to create less overspray and thus less waste. The end result was an improved system, using less paint and at a lower cost, with the savings shared by the two companies.

In addition to the environmental and cost savings, this new program enhanced DuPont's position as a supplier to Ford by creating a partnership. It would be more difficult to change a supplier of painted cars than a supplier of paint.

SELL FLOOR COMFORT AND AESTHETICS

Interface, one of the largest manufacturer and supplier of commercial carpeting, introduced a new model for providing carpeting to its customers. Instead of selling carpeting, it sold a floor covering service. Its customers pay for the comfort and aesthetics of a floor covering and pay for it on a monthly or yearly basis. Interface manufactures the carpeting in the form of tiles, usually two feet by two feet, and after it is installed, the company provides a service to maintain it. After all, the carpeting itself belongs to Interface, not the customer. Since the carpeting wears out in the major traffic areas and not along walls or under tables, only the worn carpet tiles need to be replaced. The old carpeting is returned to the manufacturing facility, separated into its original components, and recycled to produce new carpeting. With this business model, less carpeting is produced and much less is disposed of in the landfills. Again, less natural resources and energy are required for this business model.

SELL WATER TREATMENT SERVICES

Water treatment chemical suppliers, like Nalco, Calgon, or Dearborn, sell their products to a variety of applications that require a special treatment of, usually, wastewater. One such example is the treatment of paint waste from an automobile assembly plant.

During the painting of new automobiles, the overspray from the paint guns is captured by a vertical wall of flowing water. This water flows into a large basin that resembles a six-foot deep swimming pool. Here the water is treated with chemicals that enhance the coagulation of the paint particles. Various methods are used to separate the paint from the water, but for each system, having large paint particles makes the process considerably more efficient. One system may employ a horizontal blade at the surface of the water, which sweeps the large paint particles to one end of the pool for removal. While the removal of the paint includes a certain volume of water, the quantity of water is reduced for the next step, which could be filtering. Regardless of the process employed, the water is continuously circulated throughout the system.

The chemical supplier normally provides the customer with the recommended quantity of chemical addition for the system. This recommended quantity gives assurance that the paint coagulation will occur at a maximum level. In some cases, the recommended addition may include a safety factor to be sure the system works at its maximum efficiency. At the same time, the chemical supplier is also interested in maximizing its sales while keeping its relationship with the customer.

What has occurred in many situations, the customer has now hired the chemical supplier to operate the water treatment system. The supplier now has an employee working full time at the auto assembly plant with the responsibility of maximizing the efficiency of the paint separation system while utilizing the minimum quantity of the water treatment chemicals. The treatment chemicals are now a cost center for the chemical supplier and not a profit center. It has been shown that fewer chemicals are employed for this kind of system and savings are accrued to the two parties.

UNDERUTILIZED ASSETS

One of the best opportunities for a servicizing project is to consider underutilized assets. If you think about some of your own assets that are not used as often as the manufacturer intended, they can lead, or have led, to opportunities. Of course, the manufacturer would prefer the limited use of the product as it can then sell more of it.

When movies became available on video tape and later on disks (DVDs), they were purchased to watch at the owner's convenience and as often as they wished. However, movies are not viewed very often, usually no more than once or twice. So companies like Blockbuster bought the recorded movies and rented them out so that the product was used considerably more often. This business went a step further so that a DVD is not even required. Netflix allows the customer to stream the movie through the Internet and onto the television, so a physical product is not even required.

Another example of an underutilized asset is the family automobile. If a car is driven 12,000 miles per year at an average speed of 25 miles per hour, which assumes both highway driving and neighborhood driving, it amounts to 1 hour and 20 minutes per day. This means that the car is used for a little over an hour per day and sits idle for almost 23 hours per day. What a waste! This thinking was the impetus for companies like Zipcar (www.zipcar.com) and I-Go (www.igocars.org) to purchase automobiles and allow members of their organization to use them when they need transportation. Of course, the members must reserve the car in advance, pick it up at a specific location, and return it to the same or another specific location. Most of the members that might have been two-car families usually sell one of their cars within a year after becoming a member.

This service business has gone to the next step where automobile owners provide this service themselves. In the United States, RelayRides (www.relayrides.com) is a new company consisting of people who rent their *own* car, usually to neighbors. The company provides the technology to connect the car owner with the car renter and to provide the necessary insurance. In Europe, similar companies like Buzzcar (www.buzzcar.com) and WhipCar (www.whipcar.com) allow people to rent their cars to other drivers.

The transportation service business can be extended to bicycles, the most common form of transportation in the world, other than walking. Many cities have developed bicycle rental programs that allow a person to rent a bicycle at a location, ride it to the destination, and then return it to another location within the network. However, this service has been extended to the rent-your-own concept developed with automobiles. An example is a company called "Spinlister." In early 2012, it opened up access to anyone who wanted to rent a bike, anywhere in the United States. But there was just one catch: to rent a bike, you also had to list a bike. It did that to boost inventory, but has since taken away that sort of listing pay wall to new users. Shortly after its start-up, the company came out with a whole new brand—although the service would more or less stay the same. The company changed its name "Liquid" to high-light the whole idea of creating a "liquid marketplace" for sharing people's assets—like bikes!

In 2007, two roommates decided to rent out their spare bedroom to people visiting the city for a design conference [2]. An idea was born, and the company AirBnB (www.airbnb.com) took off in 2008 during the Democratic and Republican conventions in Denver, Colorado, and St. Paul, Minnesota. When Denver ran out of hotel beds, more than 300 residents used AirBnB to offer spare rooms. Unwittingly, they had opened a new tier in the holiday accommodation industry. Similar companies have been created since then with names like Crashpadder in the United Kingdom (now part of AirBnB), Couchsurfing (www.couchsurfing.org), and i Stop Over (www.istopover.com).

To develop some new businesses, one only needs to think of some underutilized assets like church buildings, which are used primarily on weekends. Another example could be something as simple as a pizza box for carryouts. The customer may use it for as little as 10 minutes to bring the food from the restaurant to the home. The pizza is removed, and the box is thrown in the trash unless it is still clean, which allows it to be recycled—still a waste due to the energy required for recycling.

CLASS FINAL PROJECTS

During my academic career, I taught sustainable strategies for about 14 years. The final project for each student required the application of one of the strategies to the student's place of work, for a public company, or for a new start-up. The student was asked to develop the concept but not necessarily be responsible for the technical design of the system. As many of the final projects employed the servicizing strategy, the following is a sample of their presentations.

MEDICATION DELIVERY

One of the large medical equipment companies provides the delivery of medicine to a patient via an intravenous system. This allows for the introduction of the medicine directly into the bloodstream for a more effective impact on the patient. These intravenous fluids are provided in one-use plastic bags, which are eventually disposed and possibly recycled. The student, who was working for a large medical supply company, presented a system where the plastic bags could be sanitized and reused. The presentation included a Life Cycle Analysis of the reuse system versus the disposal/recycle system, and it confirmed the justification for a reuse system [3].

REFILL PERFUME SHOP

The global market for perfumes, colognes, and other fragrances is about $3 billion, and the consumer's bottle has not been employed efficiently. People dispose the container before the bottle completes its life cycle, or it just sits there in peoples' shelves. A new shop will sell only the perfume (liquid), requiring the customers to bring their old empty perfume bottles, which must be the original ones, to the shop and where it would be refilled with the same scent and original brand of the old perfume bottle. This system would have economic benefit as very few new bottles would have to be produced; would have social benefits as the consumer would achieve savings and create more jobs with more shops; and would have environmental benefits by using less resources and creating less waste [4].

LUGGAGE

If we can assume that the average person takes two weeks of vacation each year and replaces the luggage every 10 years, the luggage is used about 20 weeks before it is disposed or given away. People may have the desire to replace their luggage every 10 years as new and improved models are introduced. In the meantime, it is stored in the home for the other 50 weeks per year. This leads to the manufacture and eventual disposal of luggage that may not have been used throughout its useful life. A new company would purchase the latest design of luggage and rent them to vacationers for, say, two weeks at a time. In theory, the luggage would be used 52 weeks per year and disposed after its useful use. The company will always be offering the latest and most cost-effective luggage. There will be less luggage manufactured, meaning that fewer raw materials will be consumed. In addition, there will be less material disposed in landfills.

HOME IMPROVEMENT PAINT

Many homeowners purchase a gallon of paint when they really only need one to two quarts. However, since a gallon of paint usually costs less than two quarts and there may be a need for a little more than two quarts, paint is usually purchased by the gallon. This usually results in excess paint stored in the basement of the home. To reduce and possibly eliminate this problem, a new paint store sells paint by a volume probably greater than is needed. The consumer can then return the excess paint to the company, and the company refills the container for the next customer. Under this scenario, there is never any leftover paint that is wasted. Of course, if the consumer wishes to retain some paint for a later touch-up job, a small amount can be retained in a separate container. A challenge for the company would be how to handle special order paint colors [5].

NUTRIENT SERVICES

An agricultural chemical company sells fertilizer to farmers through its supply chain. The farmer applies the fertilizer in such quantities to be sure that the plant receives the proper amount. However, a large portion of the applied fertilizer does not reach the roots of the plant, but rather is dispersed in the soil and groundwater. Perhaps it would be more effective if the agricultural chemical company provided a nutrient service whereby it would develop a system for the nutrients to go directly to the plant roots. The quantity of fertilizer chemical consumed would be considerably less, and no unnecessary chemicals would migrate into the groundwater [6].

BABY MATTRESSES

With the fertility rate in developed countries decreasing to less than a zero growth rate, there are many families with only one or two children. This means that baby crib mattresses are probably used for considerably less than their useful life. After the baby outgrows the baby crib, the crib and mattress are stored away and possibly never used again. This proposal suggests a rental service for mattresses that can be used for, say, 10 years rather than the typical one year. By improving the quality of the mattress for a small incremental cost, the useful life could be extended to, say, 14 years for a total of seven different uses. After the mattress is returned to the company, it could be recovered for sanitary purposes and the next user would have a like-new baby mattress. Applying the number mentioned earlier, there would be one-seventh as many mattresses manufactured and one-seventh as many mattresses eventually disposed.

SEAT-GO-ROUND

Instead of selling car seats, Seat-Go-Round will lease *child transportation safety*. Parents will not have to purchase a new seat for every stage of a child's growth. Instead they will lease *child transportation safety* services for a nominal fee of less than $10 per month. As the child grows or the seat becomes worn, the parents turn in the seat for a newly refurbished one. The used seat will be inspected, cleaned, and repaired and will go back into circulation for another family to use. This will ensure that the product gets maximum usage instead of sitting idle until needed again or being disposed of after one use. This process will continue until the seat is no longer able to be safely used upon which time the seats will be disassembled and the materials will be recycled. By shifting the focus from buying the car seat to renting *child transportation safety*, this will result in a more effective use of each car seat produced [7].

BIKE HELMETS

In certain developing countries where motorized bikes are the most common form of transportation, helmets are required when riding on a main road or highway. While riding on other surfaces, not only are helmets not required, but the drivers do not even wear them. However, for that possibility of riding on a main road, the driver must have a helmet in his or her possession. That leads to every rider having a helmet in his or her possession even though the use of the helmet may be limited. The proposal suggests establishing kiosks at the entrance and exit of the main roads so that bike riders can obtain a helmet when entering the main road and returning it to a similar kiosk when exiting the main road. Less helmets would be manufactured and less helmets would go unused [8].

Q CARD FOR BETTER TRANSPORTATION

The word "transportation" elicits many images: highways, traffic, smog, parked cars, and gas stations at every corner. This is the old transportation setup. The Q Card is an all-inclusive transportation smart card that brings an indispensable tool to urban travel. Bike sharing and storage, buses, trains, car sharing, and taxicabs are how people get around. All on the same smart card and smart phone app, and priced to ensure maximum savings, the Q Card is "Transportation 2.0." The Q Card goes beyond traversing through turnstiles and opening car doors. The Q Card links to a checking and credit account, automatically reloads, and keeps you on the go. The Q Card offers variable pricing to ensure value for the customer and a full transit system. With multiple transportation choices using smart card and smart phone technology, it is now easier to not own a car in the digital age.

The Q Card offers real solutions to real transportation problems. The focus is on impediments to ridership, with an entrepreneurial eye on the serving customers' needs. Public transportation has been run according to outdated models of success and failure. Currently, over $6 is lost with every transaction on the Chicago Transit Authority (CTA). With government funds not guaranteed forever, it's obvious there is an urgent business reason to adapt and close this gap. The modus operandi of the Q Card is that sustainability, cost-effectiveness, and ergonomics all go hand in hand.

The first action of the Q Card will be a Metra Station Bike Storage Facility. This facility will prove when biking is integrated into rail transit; it becomes the fastest, cleanest, and healthiest commute possible. This will force rail operators to adapt their business models regarding bicycle integration.

Applying the Octopus Card model used in Hong Kong, the Q Card will combine all major transportation systems on the same card. The up-front cost and increased revenue will be shared between transportation and financing partners.

The Q Card will introduce variable pricing for advanced purchase of fare rides. By separating pricing from a "per-ride" basis into an "overall-use" basis, the combined transit system will allow for a single transportation ride to be cost competitive with an impulse car use. Creating and sustaining the captive customer base will be as important as any per-ride fare collection increases.

The Q Card will incorporate successful trip planning and transportation tracking software to optimize customer decision-making. The smart phone accessible software (applications or "apps") will make transportation planning, purchasing and billing on the cutting edge of being easy.. These are not your grandparents' tokens. This is the Q Card [9].

FASHION CYCLE

The fashion industry requires a significant amount of resources, uses bad labor practices, and creates a great deal of waste. The main barriers to change are customer price expectations and mass consumption behaviors. Clothing is both a basic human need and a luxury item that is often thrown away after use. US marketing focuses predominantly on either low price or brand recognition, but rarely on quality. Customers from every age, race, income, and region buy clothing. However, the style and frequency of purchase varies greatly. The fashion industry is fiercely competitive with high turnover of designers and retail stores. Barriers to market entry and exit increase with the volume of items produced. There is waste and pollution throughout the life cycle from production to disposal [10].

Fashion Cycle is a concept for a women's formal wear rental franchise that would replace small dress shops and change industry practices through shifting market demand. Rather than trying to change the nature of fashion (rapid replacement), Fashion Cycle seeks to work with the consumer's desire for the latest style at the least cost while also supporting the supplier's requirement of a higher price point to provide garments that are sustainably produced. Since most of the environmental impact of clothing production can be resolved by new technologies and practices with a higher cost, increasing the wholesale price can encourage these changes. With retailers "selling" items multiple times through rental, suppliers can charge a higher price, provided their products are made to last. In addition, waste is significantly reduced by increasing the life of the product.

Rental franchises will support and promote only sustainably produced garments. Suppliers will be required to pass a sustainability review in order to be accepted as suppliers. A tiered system will be implemented to encourage suppliers in developing countries to use sustainable practices. Franchises will be encouraged to work with local retail personnel to maintain their employment and utilize their knowledge and expertise in their regional markets. Stock will be rotated through multiple stores and regions to obtain maximum utilization prior to retirement or repurposing. Retired garments will be sold to artisans that make one-of-a-kind pieces from recycled fabric, sent back to the supplier for fiber recycling, or donated.

Fashion Cycle has a first-mover competitive advantage. In addition, although the ultimate business model is an international franchise operation, the first stage of the business will be a pilot of one store to ensure market viability prior to expansion. Sustainability is built into the mission and will be incorporated into every stage of the growth of the business.

Other examples submitted by students can provide an idea of all the possibilities of more sustainable businesses. These include eco-toys shared by children, dry-cleaning hangers and bags, wedding services, and a neighborhood share system.

REFERENCES

1. White AL, Stoughton M, Feng L. *Servicizing – The Quiet Transition to Extended Product Responsibility*. Submitted to U.S. EPA Office of Solid Waste; May 1999.
2. Lenyado B. *New York Times Travel*, July 15, 2010. Available at http://en.wikipedia.org/wiki/Airbnb. Accessed 2013 July 23.
3. Mojica S. MS in environmental management & sustainability final project. Illinois Institute of Technology – Stuart School of Business, Chicago, IL; 2006.
4. Sanguansermsri D. MS in environmental management & sustainability final project, Chicago, IL; 2007.

5. Severin K. MS in environmental management & sustainability final project. Illinois Institute of Technology – Stuart School of Business, Chicago, IL; 2010.

6. Nelson A. MS in environmental management & sustainability final project. Illinois Institute of Technology – Stuart School of Business, Chicago, IL; 2002.

7. Fox DM. MS in environmental management & sustainability final project. Illinois Institute of Technology – Stuart School of Business, Chicago, IL; 2006.

8. Nguyen U. MS in environmental management & sustainability final project. Illinois Institute of Technology – Stuart School of Business, Chicago, IL; 2006.

9. Brophy J. MS in environmental management & sustainability final project. Illinois Institute of Technology – Stuart School of Business, Chicago, IL; 2010.

10. Ingram D. MS in environmental management & sustainability final project. Illinois Institute of Technology – Stuart School of Business, Chicago, IL; 2009.

CHAPTER 7

Systems Thinking Leads to "Tunneling Through the Cost Barrier"

A simple way to introduce systems thinking is with an ancient story [1].

> Beyond Ghor, there was a city. All its inhabitants were blind. A king with his entourage arrived nearby; he brought his army and camped in the desert. He had a mighty elephant, which he used to increase the people's awe.
>
> The populace became anxious to see the elephant, and some sightless from among this blind community ran like fools to find it.
>
> As they did not even know the form or shape of the elephant, they groped sightlessly, gathering information by touching some part of it.
>
> Each thought that he knew something, because he could feel a part.
>
> The man whose hand had reached an ear said: "It is a large, rough thing, wide and broad, like a rug."
>
> And the one who had felt the trunk said: "I have the real facts about it. It is like a straight and hollow pipe, awful and destructive."
>
> The one who had felt its feet and legs said: "It is mighty and firm, like a pillar."
>
> Each had felt one part out of many. Each had perceived it wrongly.
>
> This ancient Sufi story was told to teach a simple lesson but one that we often ignore. The behavior of a system cannot be known just by knowing the element of which the system is made (p. 7).

So how do you really know whether you have a system or a group of different parts? You need to ask yourself if you can identify the parts and whether these parts have an impact on each other. In addition, you need to ask if these parts

Practical Sustainability Strategies: How to Gain a Competitive Advantage,
First Edition. Nikos Avlonas and George P. Nassos.
© 2014 John Wiley & Sons, Inc. Published 2014 by John Wiley & Sons, Inc.

act differently on their own as compared to when they are with each other. Further, does the effect over time persist in different ways?

Too often, people look at the result of a particular outcome and come to a conclusion as what one single event caused that outcome. In most cases, it is not a single cause but a set of things that are responsible. This set of things, or objects, can be called a system, and the things can be people, vehicles, cells, molecules, plants, or whatever. To form this system, these things are interconnected in such a way that over time they produce their own patterns of behavior. The system can be moved, activated, constrained, or driven by some outside forces, but the result is dependent on the characteristics of the system itself. The system must consist of three kinds of things: elements, interconnections, and a function or purpose.

SYSTEM ELEMENTS

The elements of a system are the individual objects, or things, or parts that make up the system. In an automobile, the elements are the engine, the wheels, the brakes, the steering wheel, the gas tank, and others. For a tree, the elements can be the trunk, the roots, the branches, and leaves. For a house, the elements can be the foundation, the walls, the roof, water pipes, a furnace, and others.

While elements are easily recognized as being tangible objects, the elements can also be intangible. If one considers a church as a system, some of the elements can be the faith of the members or service provided by the members to the church.

Each of the elements of a system can be described as a system itself. The engine of an automobile consists of the engine block, cylinders, spark plugs, and wires. These are all elements of the engine. A leaf that is an element of the tree can be considered a system consisting of a stem, veins, tip, and blade. A furnace, an element of a house, can be described as a system consisting of a gas burner, a blower, and a thermostat. Each of these sub-elements can be broken down to sub-sub-elements.

SYSTEM INTERCONNECTIONS

The interconnections of a system describe the relationships between the elements. In an automobile, the engine operates in order to turn the wheels, while the brakes are needed in order to stop the movement of the wheels. The gas tank contains the necessary fuel that is required for the engine to operate, while the steering wheel is necessary to guide the directions of the wheels.

If one looks at the engine element as a system, it consists of the various sub-elements. The engine block is the physical part that contains the cylinders, spark plugs, and cooling water. The cylinders are needed to move in two directions and are able to transfer the movement to the wheels so that they can turn. The spark plugs are made to spark at a certain frequency to ignite the gasoline, thus making the cylinders move in the two directions.

In a tree, nutrients are absorbed from the soil by the roots and are transferred up through the trunk of the tree. The nutrients are then transferred to the branches that in turn provide the nutrients to the leaves. Thus, the leaves become larger, provide cover for the ground and objects beneath it, and absorb carbon dioxide.

In a church, the faith of the members mean that they will attend the church services, extend their faith to other potential parishioners, and provide some services to the church. The service provided may enhance the faith of other parishioners and possibly provide assistance to those in need.

SYSTEM FUNCTION OR PURPOSE

Once the elements are identified and the interconnections are described, there must be a function or purpose for the system. In an automobile, the elements and their interconnections have one primary purpose, and that is to move one or more passengers from one location to another location. This is the primary purpose, but a secondary purpose might be to attach a trailer to the rear of the automobile in order to transport a boat.

The interconnections between the elements of a tree can be for multiple purposes or functions. In this system, the purpose could be to grow the tree as large as possible in order to generate wood for construction purposes. Another purpose for the tree is to provide shade to a house and thus reduce the cost of cooling. A third purpose might be to have the tree leaves absorb carbon dioxide in order to reduce the concentration of this gaseous emission.

For a house, there are many elements and many interconnections. These are all necessary to provide shelter for one or more people. In addition to providing shelter, the purpose also includes a place to sleep, a place to eat, a place to read and write, and a place to relax.

When describing intangible elements along with the interconnections, a function or purpose is still necessary. In the case of the faith and service elements for parishioners of a church, the purpose could be to provide a place for people to worship and obtain comfort. The purpose could also be to spread the word of the religion and thus increase the number of its faithful.

While the terms "function" and "purpose" are used interchangeably, there may be fine distinction between the two. Purpose usually pertains to a human system such as the faith of a religion. Function usually pertains to a nonhuman system such as the movement of an automobile. However, these two words can be used interchangeably without much loss in the discussion of systems.

The description of a system along with its components—elements, interconnections, and functions—is described here in very basic terms. A more thorough presentation of systems thinking has been provided by Donella Meadows [1].

TUNNELING THROUGH THE COST BARRIER

Many times when people evaluate an operation to determine how to make it more efficient, the first task is to find that improvement that makes the most impact at the least cost. In other words, look for the best return on the investment.

Once that task is accomplished, people then look for another modification that could result in additional improvements. Most likely, this one will cost a little more and the result may be a little less than the first one. This process is likely to be continued until the improvements become less and less while the cost of implementation continues to increase. You reach the point of diminishing returns, and that is where you stop on trying to improve the efficiency. However, if one looks at this project as a system, there are opportunities to save even more energy, which will allow to "tunnel through the cost barrier," making the cost come down and the return on investment go up. Basically, one can make a detour to a considerably higher return on investment by considering the entire system.

An example of this is the design of the heating, ventilation, and air conditioning of a home. A builder will tell you that more efficient materials will cost much more than the normal materials that are usually less efficient. This can refer to the wall thickness, the insulation, the quality of the windows, and the efficiency of the furnace. If you consider each of these alternatives separately, you will likely follow the curve of diminishing returns. Each improvement reduces the energy consumption, but eventually you get to the point of diminishing returns when the next improvement provides little if any energy savings.

However, if you look at the whole system, the addition of better insulation and better windows will reduce the energy requirement. Now the builder can install a much smaller furnace and a smaller air conditioner. The additional cost of the insulation and windows can often result in overall savings because of the smaller furnace and A/C. Spending a little more money on compact fluorescent light bulbs for the house will not only save on the energy cost, but they will emit less heat. This will result in energy savings from the air conditioning unit [2].

INTERFACE PIPE DESIGN

Another example of "tunneling through the cost barrier" can be seen at the industrial level. Several years ago, the largest American commercial carpeting company, Interface, was building a new factory in Shanghai. It hired an engineering company to design the industrial process that required 14 pumps. In optimizing the design, a top engineering firm sized those pumps to total 95 horsepower. The design included the smallest-diameter pipes that could transfer the fluids effectively as well as the most efficient pumps. However, an Interface engineer decided that he would review the design and see if it could be made more efficient. He looked at the entire process as a system rather than optimize each component separately.

The new specifications required two changes in design. First, the Interface engineer chose to deploy big pipes and small pumps instead of the original design's small pipes and big pumps. Friction falls as nearly the fifth power of pipe diameter, so making the pipes 50% fatter reduces their friction by 86%. The system then needs less pumping energy—and smaller pumps and motors to push against the friction. If the solution is this easy, why weren't the pipes originally specified to be big enough? Because of a small but important blind spot: traditional optimization compares the cost of fatter pipe with only the value of the saved pumping energy. This comparison ignores the size, and hence the capital cost, of the equipment—pump, motor, motor-drive circuits, and electrical supply components—needed to combat the pipe friction.

The engineer did not have to calculate how quickly the savings could repay the extra up-front cost of the fatter pipe, because capital cost would fall more for the pumping and drive equipment than it would rise for the pipe, making the efficient system as a whole cheaper to construct.

Second, the engineer laid out the pipes first and then installed the equipment, in reverse order from how pumping systems are conventionally installed. Normally, equipment is put in some convenient and arbitrary spot, and the pipe fitter is then instructed to connect point A to point B. The pipe often has to go through all sorts of twists and turns to hook up equipment that's too far apart, turned the wrong way, mounted at the wrong height, and separated by other devices installed in between. The extra bends and the extra length make friction in the system about three- to sixfold higher than it should be. The pipe fitters don't mind the extra work: they're paid by the hour, they mark up the pipe and fittings, and they won't have to pay the pumps' capital or operating costs.

By laying out the pipes before placing the equipment that the pipes connect, Interface was able to make the pipes short and straight rather than long and with many turns. That enabled them to exploit their lower friction by making the pumps, motors, inverters, and other electrical components even smaller and cheaper. The fatter pipes and cleaner layout yielded not only 92% lower pumping energy at a lower total capital cost but also simpler and faster construction, less use of floor space, more reliable operation, easier maintenance, and better performance. As an added bonus, easier thermal insulation of the straighter pipes saved an additional 70 kilowatts of heat loss, enough to avoid burning about a pound of coal every two minutes, with a three-month payback.

The Interface engineering department marveled at how they could have overlooked such simple opportunities for decades. The redesign required, as inventor Edwin Land used to say, "not so much having a new idea as stopping having an old idea." The old idea was to "optimize" only part of the system, the pipes, against only one parameter—pumping energy. The engineer, in contrast, optimized the whole system for multiple benefits—pumping energy expended plus capital cost saved. Such whole-system life cycle costing, in which all benefits are properly taken into account over the long run, is widely accepted in principle but almost always ignored in practice. Instead, single components are usually considered in isolation. You can actually make a system less efficient while making each of its parts more efficient, simply by not properly linking up those components. If they're not designed to work with one another, they'll tend to work against one another.

REDUCING OIL IMPORTS

The United States has been importing oil for many years and will probably continue to do so for many more to come. While it has always been a major cost to US citizens, there have been some attempts to alleviate the situation.

In the early 1970s, there was a major shortage of oil production in the United States. Automobile owners were limited to the amount of gasoline that could be purchased at any one time. Consequently, there were long lines at gas stations, sometimes even reaching one-half block. This was primarily due to an oil embargo by the Arab members of the Organization of the Petroleum Exporting Countries (OPEC). The embargo was instituted because of the US support for Israel. About the same time, US oil production peaked as the country had depleted 50% of its oil reserves [3]. This made the United States more concerned about oil imports as US production would start to decrease. The country was concerned about increasing oil imports while an oil embargo took effect.

Several years later as oil imports continued to increase, then President Jimmy Carter, who had the ability to think in systems, suggested that we impose a tax on gasoline that would be proportional to the fraction of US oil consumption that had to be imported. If imported oil consumption continued to rise, the tax on gasoline would rise in order to reduce the demand for gasoline. The tax would be used to develop alternatives to the imported oil in order to reduce its consumption in the United States. Once the imported oil quantity fell to zero, the tax would fall to zero. Obviously, this tax never passed, primarily because President Carter could not explain it to the press and public that did not understand systems thinking.

REDUCING CARBON DIOXIDE EMISSIONS

If we think of climate change as a system consisting of all the sources of carbon dioxide emissions (elements), the impact of the emissions on our atmosphere (interconnections), and the warming of the planet (purpose), we can then suggest means of impacting the system in a beneficial way. In this particular example, the purpose or function is not necessarily a goal that we wish to achieve. Here we want to find a means to reduce the function of this system.

To achieve the goal of reducing the purpose, we must reduce the elements. However, the elements themselves can be considered the function of a system. A power plant can be considered a system with elements being the fuel, water/steam, generators, etc. The interconnections are all the controls that eventually lead to the production of electricity, which is the purpose or function. However, the production of carbon dioxide is also a "purpose," albeit a negative one. So, one way of reducing the carbon dioxide emissions is to find a system that allows for maximizing one purpose (electricity) while minimizing another purpose (carbon dioxide). This requirement can be achieved with renewable energy like solar, wind, or geothermal.

Another approach is to reduce the generation of electricity in order to achieve a corresponding reduction in carbon dioxide emissions. This can be done by imposing a tax on the carbon dioxide, which should reduce the demand for electricity. This would also apply to the carbon dioxide emitted from internal combustion vehicles that combust gasoline. This approach is not dissimilar to imposing a tax on imported oil as described earlier.

ILLEGAL IMMIGRATION

During his term of office, President Carter was also concerned with the large number of illegal immigrants from Mexico. While this problem seems to be of major concern today, it even existed in the late 1970s. Carter believed that nothing could be done about that immigration as long as there was a large gap in the living standards and working opportunities between the United States and Mexico. Keep in mind that the US–Mexican border is the only border in the world between a first-world and third-world country. Carter went on to suggest that rather than spending money on border guards, barriers, and/or fences, we should spend the money helping to build the Mexican economy. In addition, he said that this should be done until the illegal immigration stopped—when there no longer would be this great desire for Mexicans to come to the United States. This didn't happen either.

Again President Carter was thinking of this problem as a system. It consists of Mexican people (elements), means and barriers to immigrate to the United States (interconnections), and the need to have a better life (purpose). Instead of looking at the interconnections, we should be looking at the purpose. By improving the purpose—a better Mexican life—the rest of the system can function without a problem.

A good example where this system worked was in Greece. After World War II, Greece was an impoverished nation, and for the next few decades, the Greek people were coming to the United States any way possible—legally, illegally, through Canada, visiting and not returning, etc. But after Greece's economy improved and the country eventually became a member of the European Union and later joined the Euro Zone, immigration to the United States from Greece has slowed downed considerably. The recent financial crisis, however, may renew the emigration of the Greek people.

SAILBOAT DESIGN

There are situations where systems thinking is applied to a project that results in a minimum application. One such example is provided by Donella Meadows. Once upon a time, people raced sailboats not for millions of dollars or for national glory, but just for the fun of it.

They raced the boats they already had for normal purposes, boats that were designed for fishing, transporting goods, or sailing around on weekends.

It quickly was observed that races are more interesting if the competitors are roughly equal in speed and maneuverability. So rules evolved, that defined various classes of boat by length and sail area and other parameters, and that restricted races to competitors of the same class.

Soon boats were being designed not for normal sailing, but for winning races within the categories defined by the rules. They squeezed the last possible burst of speed out of a square inch of sail, or the lightest possible load out of a standard-sized rudder. These boasts were strange looking and strange handling, not at all the sort of boat you would want to take out fishing or for a Sunday sail. As the races became more serious, the rules became stricter and the boat designs more bizarre.

Now racing sailboats are extremely fast, highly responsive, and nearly unseaworthy. They need athletic and expert crews to manage them. No one would think of using an America's Cup yacht for any purpose other than racing within the rules. The boats are so optimized around the present rules that they have lost all resilience. Any change in the rules would render them useless.

In this case, systems thinking was applied to such an extent that it optimized the project extremely well. However, in application, the end result was a system with limited use.

Additional examples of systems thinking and tunneling through the cost barrier can be found in *Thinking in Systems: A Primer* by Donella Meadows or *Natural Capitalism: Creating the Next Industrial Revolution* by Amory Lovins, L. Hunter Lovins, and Paul Hawken.

REFERENCES

1. Meadows D. *Thinking in Systems: A Primer*. White River Junction, VT: Chelsea Green Publishing; 2008.
2. Lovins A, Lovins LH, Hawken P. *Natural Capitalism: Creating the Next Industrial Revolution*. Boston, MA: Little, Brown and Company; 1999.
3. Available at http://www.resilience.org/primer. Accessed 2013 Jul 1.

CHAPTER 8

Environmental Innovation Through Biomimicry

In the 1600s, Europe was hit with a crippling shortage. People had to deal with the fact that a valuable commodity was increasingly in short supply. What was it? Rags.

Rags were used to make paper, and paper was in great demand. Publishers of books, newspapers, and political pamphlets all clamored for more paper. But there just weren't enough rags. Advertisements appeared, asking women to "save their rags." In 1966, England banned the use of cotton and linen for the burial of the dead, decreeing they must be saved for making paper. One entrepreneur even suggested using the cloth from Egyptian mummies. The scarcity of rags led to fearful paper shortages in Europe and America.

Then a French scientist took a walk in the woods. Réne-Antoine Ferchault de Réaumur was an accomplished physicist and chemist. He was also a man who loved bugs. Walking in the woods one day, he came upon an abandoned wasp nest. Delighted, he began to examine it in detail, and an astounding fact dawned on him: the nest was made of paper, paper made by wasps, paper made without the use of rags. How? The wasps did it by chewing wood and plant fibers.

What wasps could do, he argued, man could find a way to do also. It took decades, but his discovery was the spark that inspired inventors to develop ways to make paper from wood pulp. Thanks to Réaumur's nature walk, we can now do what once would have been considered almost criminal: crumple up a piece of paper and throw it out [1].

Another interesting example is the process of photosynthesis, which means "using light to put something together." Here plants, algae, and bacteria take carbon dioxide, water, and sunlight to produce energy-rich sugars while releasing oxygen. Animals, including humans, take that oxygen and the

Practical Sustainability Strategies: How to Gain a Competitive Advantage, First Edition. Nikos Avlonas and George P. Nassos.
© 2014 John Wiley & Sons, Inc. Published 2014 by John Wiley & Sons, Inc.

sugars and transform them back to carbon dioxide, water, and energy. Without the sunlight, this photosynthesis reaction could not take place.

Almost 100 years ago, an Italian chemistry professor, Giacomo Ciamician, wrote in *Science* [2] magazine that one day our landscape would change from power plants to a large array of glass that would mimic photosynthesis and produce another form of energy. Through his inspiration, we now have acres and acres of solar cells made of silicon, a material that allows for the production of energy, even though it is nowhere found in the plant structure.

This is just one of hundreds of examples of innovations inspired by nature. In the classic book *Biomimicry* [3], Janine Benyus defines biomimicry in three different terms:

1. *Nature as model*   Biomimicry is a new science that studies nature's models and then imitates or takes inspiration from these designs and processes to solve human problems, that is, a solar cell inspired by a leaf.
2. *Nature as measure*   Biomimicry uses an ecological standard to judge the "rightness" of our innovations. After 3.8 billion years of evolution, nature has learned what works, what is appropriate, and what lasts.
3. *Nature as mentor*   Biomimicry is a new way of viewing and valuing nature. It introduces an era based not on what we can *extract* from the natural world, but on what we can *learn* from it.

What follows are examples of how nature has inspired or can inspire innovation that will lead to sustainable systems. Perhaps a better word than biomimicry for these new systems is "bio-inspired."

ABALONE

The abalone (Fig. 8.1) is a shellfish that is known for its delicious meat as well as its smooth inner coating that makes the shell as "hard as nails." A car could drive over an abalone shell and have no impact. It is stronger than any known ceramic, but why? It consists of an intricate crystal architecture that allows it to shrug off stress. If you can see the structure from the side, it consists of hexagonal disks of calcium carbonate stacked in a brick-wall formation. Between the "bricks" is a polymer that gives the formation some flexibility to ward off the head-on stress. The ceramics that are produced commercially, such as glass, porcelain, and bricks, are manufactured by taking earthy inorganic particles and subjecting them to heat and pressure. The result is a very strong material, but brittle and subject to cracking. Scientists

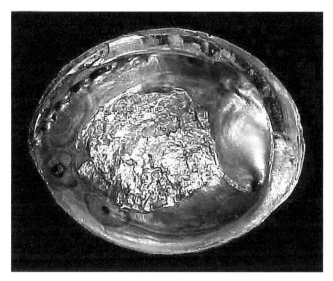

FIGURE 8.1 Abalone shell—stronger than ceramic.

are now looking at natural designs like the abalone inner coating to determine how to reproduce that structure synthetically.

Just imagine trying to build a structure that could be as strong as the abalone shell. Suppose you wanted to build a very strong structure such as a house or small office, and after it is designed, structural forms are used connected to create the walls of the building. These forms are unique as they can contain liquids such as seawater. Suppose these forms are then filled with seawater and the secret chemical used by the abalone to create its shell is added to the seawater in the forms. After a short period of time, the water is drained, and what remains are walls stronger than any ceramic or brick currently made. All that is needed to accomplish this feat is to learn how the abalone makes its shell.

SPIDER SILK

Another example is how spiders produce several different kinds of silk for various functions, such as forming a web or rappelling from drop-offs. Each one is mixed in its own gland, extruded through its own spinneret, and endowed with its own chemical and physical properties. The properties of these silks are really astounding when compared to man-made materials. Compared on an equal weight basis, some silks are five times stronger than steel and five times tougher than Kevlar, the material used for bulletproof vests. At the same time, the silk can be very elastic and stretch up to 40% of its original length, something not possible by steel wire. Just imagine if someone could learn to do what the spider does, taking a renewable, soluble material and making an extremely strong water-insoluble fiber using very little energy and generating no toxic waste. By analyzing the spider's process and reproducing it, the entire fiber industry would change dramatically.

Today, research is being conducted to produce spider silk synthetically by creating the same proteins that spiders produce. While spider silk proteins are converted to threads in nature, these same proteins can be used to prepare other morphologies, such as films, gels, foams, capsules, nonwovens, spheres, and nanofibers using different processing methods [4].

Although spider silk is already tougher and lighter than steel, scientists have made it three times stronger by adding small amounts of metal. The technique may be useful for manufacturing super-tough textiles and high-tech medical materials, including artificial bones and tendons. Researchers at the Max Planck Institute of Microstructure Physics have found that adding zinc, titanium, or aluminum in a length of spider silk made it more resistant to breaking or deforming [5]. This idea was inspired by studies showing that there are traces of metals in the toughest parts of some insect body parts. For example, the jaws of leafcutter ants and locusts both contain high levels of zinc, making them particularly stiff and hard.

BIVALVES

Bivalves is the name given to the seafood consisting of oysters, clams, mussels, and others. These creatures are usually found connected to some other object rather than just floating in the water. An amazing natural phenomenon is how these bivalves, or mussels in particular, can attach themselves to almost anything in water. If you look closely at a mussel, you can see hundreds of small translucent threads (about two centimeters long) extending like plastic tethers from the bivalve to the attached object. These tethers are called byssus, and they are truly amazing. When the bivalve wants to attach itself to an object, it will stick out a fleshy foot and create a chemical complex at the end of the byssus where there is a small disk. Through the small disk, called a plaque, a natural adhesive is squirted between the plaque and the attaching surface. Within three to four minutes, the entire process is complete, including the curing of the adhesive.

At home or at the office, anytime we wish to use an adhesive to join two objects, both objects must be dry or the adhesive may not perform as expected. The adhesives industry has been trying for many years to find an adhesive that will function in moist conditions and stick to anything. The mussels can do just that.

In 2010, scientists at the University of Chicago applied for a patent for a synthetic version of mussel glue, an adhesive, self-healing gel [6]. The potential remains exciting; biomedical transplants and repairs along with underwater mechanical engineering remain the chief areas where this material will have a significant impact. The innovation illustrates how important cross collaboration over years of research can be.

Key to their breakthrough was using metal ions for molecular bonding rather than the typical permanent covalent bonding in most synthetic polymers. Other synthetic polymers had not performed well for both strength and ductility in the past. The tightly bonded material became too brittle as it was strengthened. As any structural engineer knows, a part of a bridge that is too strong can weaken the whole structure as much as a weak part. Changing the pH of the material also enabled them to change its properties. One of the bio-innovation is obtaining strength from shape even if it is at the molecular scale.

RHINOCEROS

The rhinoceros population in the world is dwindling very rapidly—like the black rhino population in Africa, which has gone from several hundred thousand around 1900 to 65,000 in 1970 and to around 2400 in 2004. As of November 2001, this species has been officially declared extinct.

The unfortunate reason for this decline is the illegal killing of these animals for the horns protruding from their head (Fig. 8.2). Because of their composition, these horns sold for about $90,000 per kilogram in late 2011, or twice the value of gold [7]. The primary reason for this value is the booming demand in China for perceived medicinal and aphrodisiac qualities. They are also coveted in Yemen for making dagger handles.

FIGURE 8.2   Rhinoceros horn—made of keratin.

One way to stop the slaughter of rhinos is to find a synthetic substitute for the horns. When analyzed, it was determined that these horns are made of keratin, the same tough, fibrous protein that is in our fingernails. However, grinding up our cut fingernails won't do any good because it is not the keratin itself that gives the horns its coveted strength and luster, but rather the structure. Upon careful examination of a horn cross section, the pattern was recognized to be similar to the graphite-fiber-reinforced composite that has been recently engineered for race cars, tennis rackets, and the new lightweight Boeing 787 airplane. We just learned how to produce a very strong and lightweight material, whereas nature has been doing it for millions of years. However, there is still more to learn as most engineered composites do not have both compression and torsion strength like the rhino's horn—and they can't heal themselves.

FISH-INSPIRED TRAVEL

Nissan Motors was interested in developing an automobile that can be driven in congested roads without the fear of a collision. So the company looked at the behavioral patterns of a school of fish that avoid obstacles without colliding with each other (Fig. 8.3).

To develop this new automobile, Nissan noted that fish recognize the surroundings based on lateral-line sense and sense of sight and form schools based on three behavior rules (Fig. 8.4): (i) change traveling direction without colliding with other fish, (ii) travel side by side with other fish while keeping a certain distance between each fish (to match the speed), and (iii) gain closer proximity to other fish that are at a distance from them.

The company then applied laser range finder technology, a sensor that measures the distance to an obstacle with reflection of laser light, to allow the cars to travel side by side. It also applied ultra wide band, short-distance radio communications technology that measures the position of the target and the distance to it through calculation of the time lag from transmission to reflection of the pulse signal. This technology assured the cars from colliding with an obstacle.

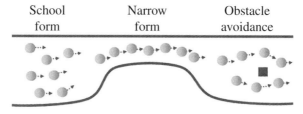

FIGURE 8.3 Fish behavior avoid collisions.

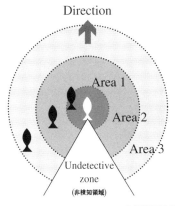

Direction

Undetective zone
(非検知領域)

Area 1: Collision avoidance
Change traveling direction without colliding with other fish.
Area 2: Traveling side-by-side
Travel side-by-side with other fish while keeping a certain distance between each fish (to match the speed).
Area 3: Approaching
Gain closer proximity to other fish that are at a distance from them.

FIGURE 8.4   Fish behavior rules.

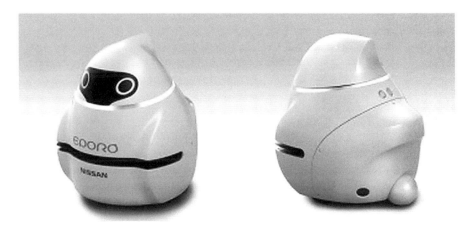

FIGURE 8.5   Nissan's robot car inspired by fish.

Through the bio-inspiration of a school of fish, Nissan has developed a robot car (Fig. 8.5) called the EPORO, an abbreviation of EPisode O (Zero) RObot (episode aiming to be CO_2-free and accident free).

CHEETAHS

Boston Dynamics, a private company with funding from the US Department of Defense DARPA Maximum Mobility and Manipulation (M3) program, has made and tested the world's fastest land robot, a four-legged machine based on the African cat (Fig. 8.6). Critical to the performance was the mimicking of the flexible spine of the cheetah, which allows for hyperextension and the coordination of the striding legs.

FIGURE 8.6 Cheetah inspiring fast robots.

FIGURE 8.7 Fastest robot inspired by cheetah.

The machine (Fig. 8.7) recently set the land speed record for robots at 28.3 meters per hour, faster than the world's fastest human (Usain Bolt at 27.8 meters per hour over 100 yards in 2009). The robot is not autonomous: an off-board hydraulic pump and a boom provide power and assist in making it steady as it runs on a treadmill. A free-running version, powered by a gasoline engine, is in the planning. When it does begin trials, it will still be far behind its natural mentor, which can reach speeds of up to 70 meters per hour. All-terrain robots that are fast and reliable would be a distinct tactical advantage to military ground forces and could be used for other hazardous work like disaster relief and medical rescue [8].

COMPACT AND EFFICIENT STRUCTURE

Many years ago, engineers were challenged to design a structural part that must be very rigid yet very light. The real design challenge was to employ the least amount of material to keep the weight very low. Perhaps the engineer was required to design, for example, an airplane wing. The goal is to develop an internal structure that could be covered with a skin which would make the wing very rigid yet very light.

If the engineer starts with a geometric design that utilizes the least amount of material for a given area, that geometry would be a circle. We know from geometry that the largest area for a given perimeter is in the form of a circle. However, if that design is a matrix of circles, there is a large amount of material where the circles are attached (Fig. 8.8). The engineer could then use the simplest geometric figure, which is the triangle, but even in this matrix there is a large amount of material at the vertex of each triangle (Fig. 8.9). Geometrically, the engineer would then go to the next design in the hierarch, which would be a square. This design would show an improvement, but there is still too much material at the intersection of four lines (Fig. 8.10). The next geometric design in the hierarchy is a pentagon, but the asymmetry makes it impossible to consider (Fig. 8.11).

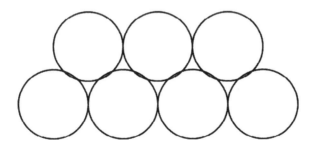

FIGURE 8.8   Structure with circles.

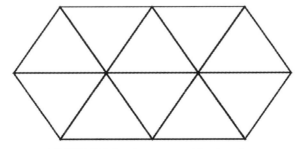

FIGURE 8.9   Structure with triangles.

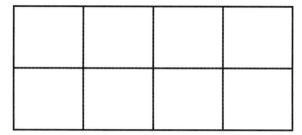

FIGURE 8.10 Structure with squares.

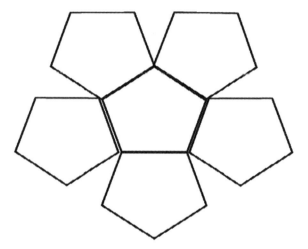

FIGURE 8.11 Structure with pentagons.

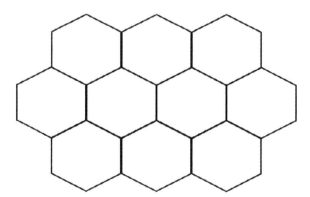

FIGURE 8.12 Structure with hexagons.

Of course, the next step would be to employ a hexagonal type structure that consists of the least number of lines at an intersection, and thus this would be the design of choice (Fig. 8.12). While this may be a design created by engineers,

FIGURE 8.13 Beehive—most efficient structure.

perhaps 100 years ago, bees have been using this same design for thousands of years in building their homes—the beehive (Fig. 8.13). Bees learned how to construct the most rigid structure with the least quantity of material.

ETHANOL

During the past 10 years, the production of ethanol as a fuel has become a very important process. For many years, up to 10% ethanol has been added to gasoline in certain geographic areas to reduce emissions. Today, ethanol is the primary fuel for flex-fuel automobiles that can use up to 85% ethanol. In Brazil, ethanol is produced from sugarcane bagasse, but in the United States it is produced from corn.

Conversion of corn to ethanol consists of a chemical process whereby enzymes are added to convert the starch in the corn to a sugar. Yeast is then added and the sugar is converted to ethanol. The process may not be very efficient as some studies report that it takes as much as 1.29 units of energy during the process to produce 1.0 units of energy as ethanol [9]. On the other hand, other studies show just the reverse, as the output–input energy ratio is 1.34 [10]. The difference in these two analyses has to do with the assumptions, primarily the allocation of the energy consumption to the by-products.

In any event, researchers have looked at ways to increase the efficiency of the conversion process. Currently, the corn-to-ethanol process utilizes only the corn kernel, as this is the only part of the corn plant that can be converted to sugar and alcohol. Scientists at Michigan State University noted that the enzyme that allows a cow to digest grasses and other plant fibers can be used to turn other plant fibers into simple sugars. They then discovered a way to grow corn plants that contain this enzyme. They have inserted a gene from a bacterium that lives in a cow's stomach into a corn plant. Now, the sugars locked up in the plant's leaves and stalk can be converted into usable sugar without expensive synthetic chemicals [11]. This new approach should make the corn-to-ethanol process considerably more efficient, thanks to a cow's stomach.

MOBILE PHONE SCREEN

As mobile phones become more and more popular and as the phone screens become larger and larger, the dependency on these phones also becomes more critical. People rely on their phones almost the entire waking day no matter where they are located to make calls, browse the Web, send text messages or e-mails, play games, watch videos, or play music. But what happens if the person wants to read the phone screen during the day under a bright sun, perhaps at a beach. It is difficult to read anything on the screen.

Qualcomm, a large American electronics firm that manufactures components for most mobile phone producers, has developed a new technology that mimics the way nature gives bright color to butterfly wings. In butterflies, a colorless, translucent membrane, covered by a layer of microscopic scales, is hit with sunlight. Light hitting the wing interacts with light reflecting off the wing, much like a prism. Light passes through the membrane, and as the butterfly's wings flap, sunlight refracts at different wavelengths. The waves of light turn color on and off, giving the wing its shimmering iridescence.

Qualcomm's screen, called Mirasol, replicates the effect with two glass panels and tiny mirrors that reflect colors onto the screen. The play of light on the mirrors defines the colors in individual cells. The device delivers bright color in strong light, only drawing power when the display is refreshed. Unlike an LCD screen, however, it doesn't work in a dark room so an e-reader would need a backlight [12].

Further development of the Mirasol technology has resulted in some very interesting benefits.

- Crisp content in bright sunlight—As a color reflective display, Mirasol is illuminated by the ambient conditions where it is used. This means that even in the brightest conditions, a Mirasol display will remain crisp and clear to view. For instances where no ambient light exists, Qualcomm has developed an integrated reading light to illuminate the display.

FIGURE 8.14 Viewing LCD display in the sun.

- Energy efficient for long battery life—The revolutionary Mirasol displays use microscopic mirrors to reflect ambient light. This means Mirasol displays do not use a backlight, one of the largest consumers of power in today's backlit displays. Mirasol displays are also bistable, which allows for near-zero power usage in situations where the display image is unchanged. This means that Mirasol displays benefit from considerable power savings, especially compared to displays that continually refresh, such as LCDs.
- Rapid refresh rate supports video—Mirasol displays are able to meet modern desires of dynamic, interactive content. With switching speed in the tens of microseconds, Mirasol displays are capable of video.

As a result of this technology being inspired by the flapping of butterfly wings, people can now read mobile phone screens or tablet readers (Fig. 8.14) in sunlight.

CRICKET SOUNDS

If you think the sound of crickets is an essential part of any romantic summer evening, you have good reason. Male crickets are trying to attract females by a species-specific chirp at a fixed frequency of about 4.5 kilohertz. But how do the females find them once they hear the call? [13]

Female crickets are equipped with special receptors or tympani on their front legs that are used to compare the pressure fluctuations between right and left sides of the insect. The pressure is carried up into the cricket by means of auditory trachea. Pressure fluctuations reach each tympanum directly as well as indirectly, from the other side of the body. Phase shifts between these two sources strongly modulate the tympanum oscillations and allow the female to determine the direction of the source of potential romance.

Crickets also have many tiny hairs or cerci along their abdomens that measure pressure fluctuations, and scientists believe their purpose is to sense attacks. The sensitivity and precision of these hairs are remarkable and inspired a research group from the University of Twente in the Netherlands to mimic their operation.

The team aimed at developing a bio-inspired perception system. They attached hundreds of 0.9-millimeter-long plastic wires to sockets in silicon wafer sheets to form a receptor array. The wires rotate in their sockets as they are moved by minute air pressures and the smallest movements are registered by the flexibly suspended plate. The electric capacity of the plate changes as a result, and this measure is fed into a central computer.

Further refinement allowed measurement at an even finer level of precision by adjusting the spring stiffness of the wires electronically. This was accomplished by investigating the alternating voltage needed to relax each wire at the required moment, enabling it to be extra sensitive to the related frequencies. The effect was significant and increased sensitivity at the adjusted frequency 10-fold.

This device will be a useful prototype for technologies like hearing aids and sensors as well as measuring devices for airflow in aeronautics. The device already reaches levels of precision in measuring air pressure and particle velocity previously unmatched.

The innovation here is an example of successful scale change and use of modularity in order to improve an existing technology. The exploration of nature at the microscale has informed the improvement, and it has the potential for creating a new paradigm for sensing in several different technologies.

This is another example of developing or extending a new technology by the inspiration from nature—in this case a cricket.

BIO-INSPIRED LEDs

Similar to inspiration provided by crickets, the nighttime twinkling of fireflies has inspired scientists to modify an LED so as to make it over 50% as efficient as an original. Researchers from Belgium, France, and Canada studied the internal structure of firefly lanterns, the organs on the bioluminescent insects' abdomens that flash to attract mates. The scientists identified an unexpected pattern of jagged scales that enhanced the lanterns' glow and applied that knowledge to an LED design to create an LED overlay that mimicked the natural structure [14].

The overlay, which increased LED light extraction by up to 55%, could be easily tailored to existing diode designs to help humans light up the night while using less energy.

Apparently, fireflies create light through a chemical reaction that takes place in specialized cells called photocytes. The light is emitted through a part of the insect's exoskeleton called the cuticle. Light travels through the cuticle more slowly than it travels through air, and the mismatch means a proportion of the light is reflected back into the lantern, dimming the glow.

The unique surface geometry of some fireflies' cuticles can help minimize internal reflections, meaning more light escapes to reach the eyes of potential firefly suitors.

Using scanning electron microscopes, the researchers identified larger, misfit scales on the fireflies' cuticles. When the researchers used computer simulations to model how the structures affected light transmission, they found that the sharp edges of the jagged, misfit scales let out the most light. This finding was confirmed experimentally when the researchers observed the edges glowing the brightest when the cuticle was illuminated from below. Human-made light-emitting devices like LEDs face the same internal reflection problems as fireflies' lanterns, and researchers thought a factory-roof-shaped coating could make LEDs brighter.

To create a jagged overlay on top of a standard gallium nitride LED, it is necessary to deposit a layer of light-sensitive material on top of the LEDs and then expose sections with a laser to create the triangular factory-roof profile. To produce a more efficient LED does not require creating a new LED, but rather just coat and laser patterns on an existing LED.

LESSONS FROM LAVASA

About 60 miles southeast of Mumbai, India, a new city, Lavasa, is under development. It is a city that will consist of five small villages with populations of up to 50,000 people. The area under development will be a challenge as it is really a dense forest along steep slopes [15].

The real problem, however, is that this area is located in one of India's monsoon hot spots. It is subject to a very short but intense rainy season of about three months with total rainfall of up to 30 feet. On top of that, this rainy season is followed by months of drought. Even though this area with steep hills experiences extensive rain followed by drought, there are no signs of erosion and the water is used efficiently all year round.

A partnership between Janine Benyus and an international architectural firm, HOK, has been retained to study the ecological system of this area and to integrate it into the design of the city. In this manner, whatever gets built will perform similarly to the natural environment. They will design the roofs of the buildings to simulate the environment and re-release the monsoonal water back into the air as water vapor. The pavements will be designed to allow water to permeate into the ground, and the building foundations will grip the hillsides like the roots of native trees.

To design the roads in the city, the team examined the behavior of anthills as they are able to remain structurally sound during the worst rains of the year. What was determined is the sinuosity (behaving like a sine wave) of the anthills allows for the water to slow down and channel over the anthills. The roads will be designed to mimic the characteristics of the anthills.

These designs will assure that the natural ecosystem of that region will not be disturbed, and at the same time the city will be enhanced by mimicking this natural environment.

MINDFUL MINING: A PROPOSAL

As a final project for a class, one student applied two bio-inspired technologies to make a mining operation considerably more efficient. The following is a student's abstract of a proposal for a company like Caterpillar [16].

Introduction

Caterpillar is the world's leading manufacturer of construction and mining equipment, diesel and natural gas engines, industrial gas turbines, and related services. Caterpillar's 2006 sales and revenue was $41.517 billion with a record profit of $3.537 billion. Caterpillar provides production, support, and technology solutions for surface and underground hard-rock mining. Mining products include large mining trucks, wheel loaders, motor graders, scrapers, and track-type tractors.

Business as Usual

Mining is a very material intensive operation. Ore waste-to-metal ratio is often very high. In the case of gold, it is as high as 10,000:1. This necessitates excessive amounts of earth to be moved to extract economically viable amounts of metal. The ecological footprint is very high. Caterpillar understands the importance of environment protection and sustainability. It has formed alliances with mining companies to reduce environmental impacts of mining. However, Caterpillar's present revenue stream is directly tied to the amount of earth that gets moved. Its product line is designed to move large amounts of earth. A larger ecological footprint translates to a higher sales and revenue.

Business Unusual: Proposed Business Model

The idea being proposed here is to use the science of biomimicry to perform precision metal and mineral extraction. Precision mining eliminates the need to remove vast amounts of earth. Using a combination of microbe-inspired mining and ant-inspired micro swarm robotics, Caterpillar could possibly create a model wherein the revenue stream is dependent on the amount of pure metal and minerals extracted and not the amount of earth moved. This has the potential to radically transform the mining industry. In addition to the immense environmental benefits, this model could potentially help Caterpillar increase its revenue and market share by making other forms of mining equipment redundant.

REFERENCES

1. Beyer R. *The Greatest Science Stories Never Told: 100 Tales of Invention and Discovery to Astonish, Bewilder, & Stupefy.* New York: Harper Collins Publishers; 2009.
2. Ciamician G. The photochemistry of the future. Science 1912;36(926):385–394.
3. Benyus JM. *Biomimicry: Innovation Inspired by Nature.* New York: William Morrow and Company; 1997.
4. Elsner M, Doblhofer E, Müller S, Schacht K, Wohlrab S. *Medical Applications Group.* Germany: University of Bayreuth. Available at http://www.fiberlab.de/research/english/medicalapp_eng.html. Accessed 2013 Jul 1.
5. Lee S-L. Improving on nature. *Nature*, April 30, 2009.
6. McKeag T. The year in biomimicry (2010): mussels, elephants, water bears and more, *The Biomimicry Column*, February 24, 2011.
7. Available at http://www.freakonomics.com/2011/08/15/why-are-rhino-horns-twice-as-valuable-as-gold/. Accessed 2013 Jul 1.
8. McKeag T. The year in biomimicry (2012): robots inspired by cheetahs and moon jellies, *The Biomimicry Column*, January 9, 2013.
9. Pimentel D, Patzek T. Ethanol production using corn, switchgrass and wood: biodiesel production using soybean and sunflower. *Nat Resour Res J* 2005; 14(1):65–76.

10. Shapouri H, Duffield JA, Wang M. The energy balance of corn ethanol: an update. US Department of Agriculture, Office of the Chief Economist, Office of Energy Policy and New Uses. Agricultural Economic, Report No. 813; July 2002.

11. Sticklen M. Plant genetic engineering for biofuel production: towards affordable cellulosic ethanol Nat Rev Genet 2008; 9(6):433–443.

12. Waldrop MM. Brilliant displays, *Scientific American*, November 2007.

13. McKeag T. The year in biomimicry (2011): how beetles, mantis shrimp and more inspired innovation, *The Biomimicry Column*, January 4, 2012.

14. Energy Manager Today Staff. Bio-inspired LEDS 55% more efficient, *Energy Manager Today*, January 9, 2013.

15. Berg N. Inspired infrastructure, *ENSIA-Environmental Solutions in Action*, 2013 Jan 9.

16. Mathew M. MS in Environmental Management and Sustainability, Illinois Institute of Technology – Stuart School of Business, Chicago, IL; June 2007.

CHAPTER 9

Base of the Pyramid

In 1932, the US president, Franklin Delano Roosevelt, made a radio address titled "The Forgotten Man." As part of his address he said, "These unhappy times call for the building of plans that rest upon the forgotten, the unorganized but the indispensable units of economic power...that build from the bottom up and not from the top down, which put their faith once more in the forgotten man at the bottom of the economic pyramid." This is the first known use of the term "bottom of the pyramid," which has also been modified slightly to "base of the pyramid" (BOP).

That was over 80 years ago, but today this term refers to the billions of people in the world living on less than $2 per day. This was defined by University of Michigan Professors C.K. Prahalad and S.L. Hart in 1998. They each wrote excellent books that thoroughly described the business model and its benefits. Prahalad published *The Fortune at the Bottom of the Pyramid* [1] in 2004, while Hart published *Capitalism at the Crossroads* [2] in 2005. Hart subsequently went to the University of North Carolina and is currently at Cornell University where he heads the "Enterprise for a Sustainable World."

Prahalad proposes that businesses, governments, and donor agencies stop thinking of the poor as victims and instead start seeing them as resilient and creative entrepreneurs as well as value-demanding consumers. He proposes that there are tremendous benefits to multinational companies (MNCs) who choose to serve these markets in ways responsive to their needs. After all, the poor of today are the middle class of tomorrow. There are also poverty-reducing benefits if multinationals work with civil society organizations and local governments to create new local business models. How can this business model be employed?

One of the major environmental issues today is climate change. The United Nations' Intergovernmental Panel on Climate Change in its "4th assessment

Practical Sustainability Strategies: How to Gain a Competitive Advantage, First Edition. Nikos Avlonas and George P. Nassos.
© 2014 John Wiley & Sons, Inc. Published 2014 by John Wiley & Sons, Inc.

report" concluded, "Most of the observed increase in global average temperature since the mid-20th century is very likely due to the observed increase in anthropogenic greenhouse gas concentrations" [3]. The big question now is what can we do about it. The obvious answer is to reduce the quantity of fossil fuels for energy by employing more efficient systems. But what else can we do?

Scientists and engineers have developed processes and products that are very innovative and utilize much less energy. In some cases, however, these products are not accepted by the major markets because they would destroy an existing system or infrastructure—thus creative destruction. An example is the auto industry, which along with electric power generation is the largest contributor to global warming.

The energy required to power an automobile is used primarily to move the vehicle itself, which accounts for about 95% of the total weight. This means that only 5% of the fuel is used to move the driver from one location to another. The simple answer is to manufacture lighter automobiles. Several years ago, the Rocky Mountain Institute [4] developed an automobile made from high-tech ultralight materials, which are stronger than steel, and the resulting car can achieve over 100 miles per gallon. This technology has been rejected by Detroit because it would eliminate the assembly plants as they exist today. The investment in the auto assembly plants is too great to change to a new technology. Is there any alternative?

THE GREAT LEAP DOWNWARD

Economists have described the global economic pyramid as consisting of three layers. At the bottom, or BOP, are the 4.2 billion people earning less than $1500 per year. In the center are the 1.8 billion people in the emerging middle class earning between $1500 and $15,000 per year. At the top of the pyramid are the one billion "wealthy" people earning over $15,000 per year. Most manufacturing companies market their product to the top of the pyramid where the "wealthy" consumers exist. But should we be looking at the BOP?

Most companies market their new technologies to the top of the pyramid as this is the "wealthy" part of the global market. These technologies usually have costs associated with the development, marketing, and sale of the products that can only be accepted by the top of the pyramid. Reducing the cost of the product to reach the bottom of the pyramid will usually reduce the quality and make the product of little interest. Clay Christensen [5] talks about disruptive innovations, which are products and services that are not as good as those in the mainstream markets and therefore are of interest only to some niche nontraditional markets.

Hart and Christensen talk about the great leap downward [6] as the BOP is the ideal target for new disruptive technologies for at least two reasons. The opportunities for a new technology to penetrate a low-income market are much greater just based on the size of the market. In addition, by marketing to the BOP, a company can provide a market or service that would not otherwise be available.

An excellent example is that of the business started by Soichiro Honda of Japan [7]. In 1946, he started Honda Technical Research Laboratory, and his first project was to make use of war surplus Tohatsu and Mikuni generator motors. With bicycles being the primary mode of transportation in Japan, and other countries, he attached them to bicycles in order to provide basic transportation for this war-torn nation. Recognizing the diminishing supply of surplus motors, Honda formed Honda Motor Company, Ltd. in Hamamatsu in 1948. The company's first headquarters was a 12×18 foot shed that housed 13 employees. The company built the "A" model bicycle, the "B" model tricycle, and then the "C" model motorcycle. The company then made larger motorcycle engines and by 1955 was the largest motorcycle manufacturer in Japan. Within four years, the company became the largest motorcycle manufacturer in the world.

Honda's vision was to become an international company and that included selling its products in the United States. However, the US market for motorcycles was dominated by Harley-Davidson, which made much larger motorcycles. In addition, the market was not that large compared to the US population. In 1965, Honda produced its first automobile and four years later imported its N600 sedan to the United States. Thus, Honda went from adding motors to bicycles, and then to manufacturing motorbikes for the BOP, to selling automobiles to the entire world population.

Another example of this great leap downward is from a Chinese company called Galanz. It was formed in 1978 to sell duck feathers and subsequently became a textile manufacturer. In 1993, it entered into an agreement to manufacture microwave ovens for Toshiba. The company started slowly and only manufactured 10,000 units per year. At that time, only 2% of the Chinese households owned a microwave oven. Using Chinese materials and Chinese labor, Galanz developed and manufactured an energy-efficient microwave that was affordable to the large middle- and lower-middle-class families in China. With the volume increasing, the average price of the unit started to decrease and became affordable to more families. As a result, Galanz's domestic market share increased from 2% in 1993 to 76% by 2000 when the market was even much larger. Employing the "great leap downward" model, it started selling its products worldwide, and by 2002 it enjoyed a 35% share of the global market. Today, it produces over 15 million microwaves per year and has a 40% share of the world market [8].

ELECTRIFY THE BOTTOM OF THE PYRAMID

There are about 1.5 billion people who live without electricity, and they are at the bottom of the pyramid. Extending the grid is not always practical, and in many cases it is almost impossible due to logistics and high costs. One answer may be distributive generation where there are local, self-contained power generation and storage systems.

The importance of providing electric power generation can be explained using Figure 9.1 [9]. It correlates per capita electricity consumption with the human development index, which is a very general measure of well-being, by looking at life expectancy, literacy, education, and standard of living. Of course, as the electricity consumption of a country increases, the well-being of the people increases. Electricity provides lights, which, in turn, allow children to study in the evening, refrigeration, pumping of water into the home, power for computers and charging mobile phones, and many other benefits that enhance the people's well-being.

To provide this electricity to the bottom of the pyramid would normally require the development and installation of an electric grid as well as the

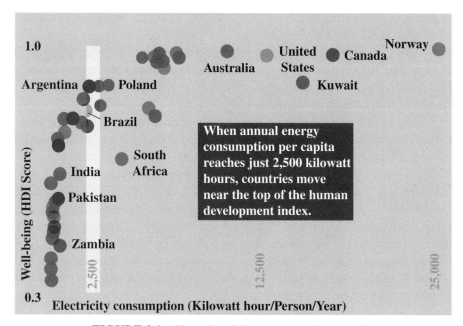

FIGURE 9.1 How electricity powers well-being [9].

generation of the electricity. The latter could be accomplished by a coal-fired power plant, a nuclear power plant, or diesel-powered generators. For remote villages, this is not the most efficient means of delivering electricity even if there was no concern with carbon emissions from the coal power plant or any safety concerns with the nuclear power plant. Stand-alone power generation systems using renewable sources like solar, wind, biomass, geothermal, or hydroelectric would make more sense given the ability to store the energy to eliminate intermittent generation when the sun is not shining or the wind is not blowing. The critical aspect of this system is storage of the energy.

The US Department of Energy, through its Advanced Research Projects Agency, continues to invest in battery research in order to develop higher-efficiency batteries at significantly lower costs. Successful development of these batteries, while primarily targeted to increase the range of electric vehicles, will create a larger market at the bottom of the pyramid. This huge market will reduce the cost of the batteries and make them available to more people. Eventually, this technology could then leap up to the top of the pyramid and develop a distributive generation system for the developing countries as proposed by Jeremy Rifkin [10].

HINDUSTAN LEVER AND NIRMA

This is an interesting case study of two companies that focused on different levels of the economic pyramid. Hindustan Lever, Ltd. (HLL), a subsidiary of Unilever, the large international British company, was serving the people at the top of the pyramid [2]. This company noted that in the 1990s, a small Indian firm, Nirma, Ltd., was offering detergent products for the BOP. Nirma had developed a business model to serve the underserved by providing a new product formulation, a low-cost manufacturing process, a wide distribution network, special packaging for daily purchasing, and pricing for consumers with limited means. The Nirma price was Rs 3 per kilogram, while the HLL price for their detergent was Rs 13 per kilogram [11].

While HLL was content in serving the top of the pyramid, Nirma grew rapidly in a market completely ignored by HLL. As Nirma grew, it started to move up the pyramid as a result of its strong base at the bottom of the pyramid. HLL saw the growth of Nirma and in 1995 responded by developing its own business model for the BOP market and diverging from its traditional business model.

HLL introduced a new detergent product called Wheel, which was reformulated for the poor people that wash their clothes in rivers and other public water systems. The company revised its production, marketing, and distribution to reach the large number of people living in the rural areas that comprise the BOP. In addition to all of these marketing changes, it also reduced the price in order to be attractive to the BOP.

Today, Nirma and HLL are still competing as they are splitting evenly about 80% of the detergent market in India. It was a very wise move for HLL to make that leap downward as the BOP market accounts for over 50% of HLL's total revenues and profits. It is even a better market for Nirma.

BOP PROTOCOL

Stuart Hart started developing the BOP strategy while a professor at the University of Michigan. He then joined the faculty at the University of North Carolina Kenan-Flagler Business School where he and other faculty formed the BOP Learning Laboratory [12]. Working with Dan Vermeer, a member of Coca-Cola's strategic "think tank," they looked at the use of anthropological approaches to understanding the needs of people living at the BOP. They determined that a systematic approach was desired by companies such as Coca-Cola that could be used to understand the needs and opportunities at the BOP. In subsequent conversations, Hart and Vermeer fleshed out the broad contours of what would become the BOP Protocol project that is now housed at Cornell University, where Hart is currently a professor.

This project moved forward by engaging five different institutions:

1. Johnson Graduate School of Management at Cornell University
2. Stephen Ross School of Business at the University of Michigan
3. World Resources Institute (WRI)
4. William Davidson Institute
5. S.C. Johnson Foundation and the Wingspread Conference Facility

and four partner companies:

1. DuPont
2. Hewlett Packard
3. S.C. Johnson, Inc.
4. Tetra Pak

This BOP Protocol continues to move forward with projects in India with Solae Company and in Kenya with S.C. Johnson. Because the people at the bottom of pyramid can be found in almost any country, the BOP Protocol even developed a project in the United States. The project was initiated in 2008 by seeking to innovate and develop new grassroots enterprises focused on improving the health and/or health care of Flint and Genesee County in the State of Michigan. By working with community resources, local partners, and poor and vulnerable families and individuals, this multiyear project would follow the BOP Protocol's participatory business innovation process to cocreate and implement sustainable, community entrepreneur-led businesses.

INITIATIVES BY THE WORLD RESOURCES INSTITUTE

The WRI is a global environmental think tank that goes beyond research to put ideas into action. It works with governments, companies, and civil society to build solutions to urgent environmental challenges. WRI's transformative ideas protect the Earth and promote development because sustainability is essential to meeting human needs and fulfilling human aspirations in the future. The organization's efforts are directed toward:

- Climate, energy, and transport
- Governance and access
- Markets and enterprise
- People and ecosystems

It applies various sustainable strategies to accomplish its goals in these various areas. One of the strategies that WRI applies is the BOP.

Jamshyd N. Godrej, a member of the WRI board of directors, was interested in bringing electric energy to the village of Rupahi in Bihar, India, which is too isolated to be connected to the grid [13]. For energy, the residents rely on whatever is available to them—kerosene, firewood, dung, or diesel generators, all of which left their houses dim and full of smoke. Installed at the community level, small hydro and biomass gasification can supply electricity to an area not covered by the grid, and solar lanterns and energy-efficient stoves can replace "dirty" fuels like kerosene and wood. In Rupahi, villagers now get their power from Husk Power Systems (HPS), a company that converts rice husk into electricity via 35–100 kilowatt mini power plants. HPS provides villages in the country's rice belt, like Rupahi, with cost-effective and environmentally friendly electricity. Providing electric energy to the rural poor people of India has been estimated to be a $2.1 billion per year market.

WRI also believes that there are investment opportunities within the BOP. *Impact investing* [14] is gaining increasing traction among a wide range of debt and equity investors, including pension funds, family offices, private wealth managers, foundations, individuals, commercial banks, and development finance institutions. Many impact investors choose to focus either in emerging markets or in developed markets. Part of the reason for this specialization is the significant regional differences that require local expertise. But another driver is investors' value sets: some prefer to help the world's poorest in emerging economies; others prioritize their local neighbors in need.

Within the developing world, the growing suite of impact investors is focusing on particular regions and sectors. Gatsby Charitable Trust and the Bill & Melinda Gates Foundation direct some of their investment capital to positively impact the lives of smallholder farmers in sub-Saharan Africa. Gray Ghost Ventures, Acumen Fund, and Omidyar Network all have programs that actively invest in alleviating poverty by financing innovations directed at India's low-income populations. In a report by The Rockefeller Foundation [15], it estimates the value of this emerging impact investing sector to be between $400 billion and $1 trillion, with profit potential between $183 billion and $667 billion over the next decade.

DEVELOPING THE BOTTOM OF THE PYRAMID

In *Capitalism at the Crossroads*, Hart describes the pathway to developing the BOP. Companies must move *beyond greening* and look at technological change and leapfrog to clean technologies that could result in disruptive business models. One of the challenges is to become indigenous and learn to codevelop technologies, products, and services with nature and local people. In this manner, the MNCs can become native to these regions.

The MNCs have been ingrained to develop strategies that take a top-down approach, and this must be altered if they wish to develop a strong market presence in the BOP. This was exemplified by Hindustan Lever, Ltd. until it learned from Nirma. Nike also demonstrated its problems when it attempted to market the World Shoe. It was attempting to market the product to the BOP, but it was still employing top-down marketing.

In many cases with respect to marketing to the BOP, companies must think in terms of creative destruction rather than continuous improvement. Companies may have developed technologies that were never introduced because they would disrupt a current situation. However, at the BOP, these same technologies could have an application as the potentially disruptive market doesn't exist. Introducing LED light bulbs to the developing countries has been very slow as the incandescent and compact fluorescent bulbs have dominated the market. However, combining LED bulbs with inexpensive solar battery chargers can provide light to areas that are not on the electric grid.

In Nike's World Shoe project, one of the failures was due to the pricing structure. The company failed to reduce the cost because it employed the same overhead as its standard products. China's Galanz was successful because it used local materials and local labor and sold to local markets. Its cost structure was such that it was able to achieve a 75% market share in its own market.

Most MNCs think about large-scale projects with large capital investments if they are going to make an impact. However, when looking at the BOP, companies must consider starting at a smaller scale working with the smaller markets and growing organically as the markets respond. Disruptive technologies are typically smaller in scale and more distributive in scale.

Finally, it is critical that the company must align the organization that will be pursuing the BOP project. Everyone involved in the project must buy into the elements of the organizational infrastructure. A misalignment can be fatal to a project as was demonstrated by Nike's World Shoe endeavor. The elements and alignment can be summarized as follows:

- Vision/Mission—setting the sustainability goal
- Goals—establishing measuring targets
- Strategy—identifying the sustainable value portfolio
- Structure—creating separate experiments, ventures, and funding
- Systems—designing new measurement, rewards, and project-evaluation tools
- Processes—enabling new technology, product, and market-development approaches
- People—integrating sustainability into recruiting, leadership development, and performance evaluation

In order to assure success, aligning these elements of organizational infrastructure should not be underestimated.

IS THE BASE OF THE PYRAMID A MIRAGE?

Before a company decides to adopt the BOP strategy, it should carefully assess the market opportunity. Although this BOP strategy was initially developed at the University of Michigan, another professor at the same school, A.Karnani, questions the validity of this strategy. He doesn't believe that it can be adopted successfully and believes that it even harms the poor people in two ways. "First, it results in too little emphasis on legal, regulatory, and social mechanisms to protect the poor who are vulnerable consumers.

Second, it results in overemphasis on microcredit and under-emphasis on fostering modern enterprises that would provide employment opportunities for the poor. More importantly, the libertarian proposition grossly under-emphasizes the critical role and responsibility of the state for poverty reduction" [16].

The developers of the BOP set the market size at $13–15 trillion, whereas Karnani believes it is really $360 billion. The major attraction of the BOP is this huge untapped purchasing power. The poor, however, may not really have the ability to save any money, so their purchasing power may be limited. The BOP strategy relies on decent margins so the company can operate profitably. The company, on the other hand, may not be able to make a profit as the BOP market is very price sensitive and the cost of serving this market can be high. In addition, the poor spend 80% of their money on food, clothing, and fuel and really don't have much remaining to purchase a "technology" product that is planned for development through the BOP. Technology products may be the only sector that could be developed at the BOP because other products would result in reduced quality if the price is reduced. Karnani also believes that small- to medium-sized companies are better suited to exploit the BOP rather than the large MNCs.

REFERENCES

1. Prahalad CK. *The Fortune at the Bottom of the Pyramid*. Upper Saddle River, NJ: Wharton School Publishing; 2004.

2. Hart SL. *Capitalism at the Crossroads: The Unlimited Business Opportunities in Solving the World's Most Difficult Problems.*Upper Saddle River, NJ: Wharton School Publishing;2005.

3. United Nations Intergovernmental Panel on Climate Change, 4th assessment report.Available at http://www.ipcc.ch/pdf/assessment-report/ar4/syr/ar4_syr.pdf. Accessed 2013 Jul 2.

4. Lovins AB, Cramer D. Hypercars, hydrogen and the automotive transition. *Int J Veh Des* 2004; 35 (1/2):50–85.

5. Christensen C. *The Innovator's Dilemma: When New Technologies Cause Great Firms to Fail*. Boston, MA: Harvard Business Press; 1997.

6. Christensen C, Hart SL. The great leap: driving innovation from the base of the pyramidsloan. Manage Rev 2002; 44(1):51–56.

7. Available at http://smokeriders.com/History/Honda_History/body_honda_history. html. Accessed 2013 Jul 22.

8. Available at http://en.wikipedia.org/wiki/Galanz. Accessed 2013 Jul 2.

9. Majumdar A. Electrify the bottom of the pyramid. *Harvard Business Review*, January–February, 2012, p. 55.

10. Rifkin J. *The Hydrogen Economy: The Creation of the Worldwide Energy Web and the Redistribution of Power on Earth*. New York: Jeremy P. Tarcher/Putman; 2002.

11. Available at http://www.moneycontrol.com/news/special-videos/the-nirma-story-how-it-washed-away-allb-school-myths_497539.html. Accessed 2013 Jul 2.

12. Available at http://www.bop-protocol.org/about/history.html. Accessed 2013 Jul 2.

13. Clean power to people. *The Economic Times*, February 28, 2011.

14. Elsen T, Gasca L. *Growing Optimism for "Impact Investing."* Washington, DC: World Resources Institute, December 7, 2010.

15. Impact investments: an emerging asset class. JP Morgan and The Rockefeller Foundation Publications, November 20, 2010.

16. Karnani A. Romanticizing the poor harms the poor. University of Michigan Ross School of Business Working Paper No. 1096, October 2007.

CHAPTER 10

Sustaining Fisheries

One definition for sustainable development, or sustainability, is meeting the needs of today's generation without compromising the ability of future generations to meet their needs. The environment consists of many different species of natural resources that are continuously being replenished, and we must develop strategies to be sure we don't deplete these natural resources faster than they are replenished. While there is a need to develop a specific strategy for each of these resources, here in this chapter, we address one specific resource—that of the many fisheries in the seas of this planet Earth. Similar strategies can be developed for other natural resources.

As the population of the earth continues to increase at a rate of about 10 million people every six weeks, or one billion people every 12 years, more and more people are migrating to the coastlines. Two resources are drawing them to the seas, water and protein in the form of seafood. As a result, humankind is in the process of annihilating coastal and ocean ecosystems. Population distribution is increasingly skewed. Recent studies have confirmed that the overwhelming bulk of humanity is concentrated along or near coasts on just 10% of the earth's land surface. As of 1998, over half the population of the planet lives and works in a coastal strip just 120 miles wide, while a full two-thirds are found within 240 miles of a coast [1].

For example, of China's 1.3 billion people, close to 60% live in 12 coastal provinces, along the Yangtze River valley, and in two coastal municipalities—Shanghai and Tianjin. With the exception of India, the bulk of Asia's population is coastal or near coastal. Of the region's collective population of about four billion, 60% live within 250 miles of a coast. No one in Japan lives more than 70 miles from the sea. Furthermore, 77% of all Japanese live in urban areas along or near the coast. About 75% of the Latin American and Caribbean population live within 120 miles from the coast. In the United

Practical Sustainability Strategies: How to Gain a Competitive Advantage,
First Edition. Nikos Avlonas and George P. Nassos.
© 2014 John Wiley & Sons, Inc. Published 2014 by John Wiley & Sons, Inc.

States, about 60% of the people live in counties that border the Atlantic and Pacific Oceans, the Gulf of Mexico, and the Great Lakes. The five states with the greatest rise in population are all coastal: California, Texas, Florida, Georgia, and Virginia. By the year 2025, nearly 75% of Americans are expected to live in coastal counties.

Another reason for a desire to live near water is biophilia, a love for living systems [2]. Depending on which historical view one accepts, human beings have been on this earth for 50,000–100,000 years. In about all but last 200 or so years, people have lived near water, in the mountains, forests, savannahs, and open fields. What has been imbedded in the genes of humans is that living in these areas is normal. During the past 100–200 years, life has transformed to living among steel, concrete, glass, and bricks. Relative to the earlier years, this is not normal so people travel to water, mountains, forests, and open fields when they want to feel good, like when they take a vacation. This is why most paintings in homes are that of landscapes. This is also the reason that lakefront homes and homes bordering golf courses are more desirable.

DEPLETION OF FISH

In the developing countries, one of the reasons that people live near water is the source of food. However, seafood is also in demand by the developed countries and by people away from the coast. The population explosion coupled with the demand for high-protein, low-fat food is contributing to the depletion of fish. In addition, other reasons include [3] the following:

1. Lack of property rights with respect to the sea. It is difficult to establish ownership of the seas.
2. There is information failure because fishermen do not know the size of available fish stocks.
3. Externalities are associated with fishing, given that fishermen fail to take into account the impact of their actions on others, including the impact of overfishing on other fishermen in the future.
4. The high fixed cost of boats is an incentive to fish as much as possible in order to cover the fixed costs.
5. Fishermen are often heavily subsidized, which encourages them to catch more.
6. There is also a prisoner's dilemma, which means there is an incentive for fishermen to catch as much as they can because that is what they expect others to do. A prisoner's dilemma is any situation where

the "payoff" from an action depends upon decisions made by other parties. This means that an individual's behavior is influenced by the predictions that they make about how others will react in response to their behavior. In this case, fishermen may predict that all other fishermen will try to catch as much as they can before the stocks run out. The combined outcome is that shrinking stocks encourage more fishing and not less! Even if fishermen agree to limit their catch, many will expect others to cheat; hence, cheating would become the "norm."

7. Finally, widespread pollution of the sea has also contributed to the gradual depletion of fish stocks.

Despite the continuing depletion of fish stocks, many fisheries continue to develop marketing strategies to increase the sale of fish. One strategy is to rename the fish species in order to increase its appeal. One of the best examples is the Patagonian toothfish, which is found in the southern Pacific, southern Atlantic, and Indian Oceans. In the early 1990s, this fish was relatively unknown in the United States, but after it was renamed Chilean sea bass, consumption started to take off. The interest was because of its oily taste and difficulty in overcooking it. However, "environmentalists warned that unless demand is reduced, the fish may face commercial extinction in as little as five years" [4]. Some regulations were imposed, but the fish continued to be overharvested, and most of the fish was caught illegally. In 2002, a campaign was initiated by restaurant chefs to take this fish off the menu. This has helped reduce the consumption, but the fish continues to be on a danger list for extinction.

This scheme to increase sales of fish by changing their names to something more appealing has continued. Other examples include bocaccio, a kind of rockfish, which is now called Pacific red snapper, or dogfish, which is now known as rock salmon [5].

OVERFISHING REMEDIES

A variety of remedies have been tried, usually in combination. These regulations included the size and type of fish that could be caught in a particular fishery, the mesh of the nets to be used, how frequently boats may go to sea in terms of days per month, and the total weight of the fish catch. In some countries, fishermen were given inducement not to fish, not unlike in the United States when farmers were once paid not to grow certain crops. One of the simple methods to increase the fish stock is just to close the area until the stock level has recovered.

In 1976, the US Congress passed the Magnuson–Stevens Fishery Conservation and Management Act (MSA), which expanded federal jurisdiction of fisheries from 12 miles to 200 miles offshore. The various regulations imposed by the MSA forced fishermen to compete with each other and with the regulators. When regulators shortened the fishing season, the fishing companies just increased their fleet size and installed more powerful engines, allowing them to cover a larger area. The regulators responded with further cuts in the season, so the fishermen put out more hooks, lines, and nets. These actions led to more cuts by the regulators. This "cat and mouse" game led to some extremely short fishing seasons. The annual commercial fishing season for Alaskan halibut was eventually reduced to 48 hours [6].

One result of this competition between regulators and fishermen is the huge increase in fishing capacity. There is a larger number of boats and associated equipment trying to catch the fish than would be necessary if the fishing process was controlled in a better manner. Not only are fishermen competing against each other and the regulators, but their capital investments are significantly higher than necessary as the resulting assets may sit idle for a better portion of the year. The same unnecessary investment would take place with the fish processors that sit idle for a portion of the year. The inevitable result is declining fish catch per boat, leading to declining revenues and increased costs.

Most of these measures have helped as the recovery of stocks in various places has been seen. For example, striped bass and North Atlantic swordfish have returned along the east coast of the United States. Other recoveries include halibut in Alaska and haddock in Georges Bank off the coast of Maine.

Government intervention is very important; otherwise, what could happen can be exemplified by the waters off the coast of Somalia, a country that has been devoid of a working government since 1991. The coastal stocks of fish are so minimal that fishermen have taken up piracy as an alternative livelihood.

Government regulations have certainly helped bring control over the fishing industry. However, these gains have come too slow and at a high price. What follows should be a better way to manage this so important natural resource.

LIMITED ACCESS PRIVILEGE PROGRAMS (LAPPs)

A new management approach for fisheries was first implemented in the 1970s in Australia, New Zealand, and Iceland [7]. This management tool was called Limited Access Privilege Programs (LAPPs) or "catch shares" for short. This system dedicates to an individual fisherman a secure share of fish. It would also dedicate a secure share to a fishery association or a community. Once that annual catch limit is determined, the catch shares represent a percentage share of annual limit. In this manner, fishermen would know exactly how much fish they would be allowed to harvest at the beginning of the season.

Similar to shares of a company stock, these catch shares can be sold or traded, thus allowing fishing companies to operate more efficiently. Each fisherman can determine the value of the catch shares to his or her own operation by comparing the catch share price with the cost of operating the fishing vessel. If shares on the market are cheaper than a fisherman's cost to catch the fish, he or she may try to purchase additional shares. On the other hand, if the cost of catching the fish is more expensive than the value of the catch shares, the fisherman will sell his or her shares and not purchase any additional ones. This system makes the fishing system more efficient and sustainable. If a fisherman is innovative and can find ways of lowering the cost of the operation, he or she will be rewarded. No longer will it be necessary for fishermen to race out to the seas to catch the maximum number of fish. The shift is to maximize value rather than volume.

> As the fishery moves towards a more efficient level, capacity is reduced and seasons expand. With a slower pace of fishing, fishermen can more effectively plan their season, reducing the amount of gear deployed, reducing bycatch, delivering fish when the market demands and staying ashore in unsafe conditions. With catch levels controlled, regulators are able to relax many of their previous effort constraints (7, p. 10).

The LAPP can provide fishermen with a management tool that maximizes the efficient use of their assets and at the same time providing for sustainable fisheries. This program results in six fish management objectives:

1. *Complying with Catch Limits*   Setting a limit on the annual catch of a species is one of the most important objectives for the management of a fishery. In order to develop a sustainable fishery, a catch limit must be determined based on the current population, and, most important, the catch limit must be enforced. Without catch limits, fishermen will land as much as possible and negatively impact the sustainability of the fishery.

   As the population of the fishery starts to increase, the catch limit can also be increased. If the fishermen remain within the catch limits, the popularity of a certain species can increase because a year-round supply can always be delivered.

If the catch limit is not enforced, the population of the fishery will continue to decrease rather than increase and make that fishery unsustainable. This is no different than working off the principal of a financial asset, a small amount each year. Eventually, the principal of that asset will disappear.

2. *Better Science and Monitoring*   It is critical to employ the best available science to determine both the population of a particular fishery and the rate of productivity of that species. Given these two data points, a catch limit can be determined to provide for a sustainable fishery. This catch limit can be increased as the population grows to sustainability. It is also critical to monitor the program to be sure the appropriate catch limit was implemented. Without an excellent monitoring system, deviations from the planned program cannot be observed.

    In 1990, the allowable catch limits for rockfish off the coast of California, Oregon, and Washington were set at relatively high levels based on incorrect assumptions. It was assumed that this species was as productive as a similar species in Alaska. The population of the rockfish started to decline rather than the planned increase. The catch limits were revised in 1993 but not imposed until 1997. During this period, the population of rockfish declined to a dangerously low level, and the cause of this decline was attributed to poor scientific data. Better monitoring leads to better science, better science leads to reduced uncertainty, and reduced uncertainty leads to more appropriate catch limits [7].

3. *Reducing Bycatch*   Bycatch is the term given to fish and other animals that are caught in addition to the targeted species but eventually discarded. This could include sea turtles, dolphins, corals, and sponges. Globally, about a fourth of the world's total catch is tossed back for commercial and regulatory reasons even though most of it is already dead or dying [8]. In some fisheries, bycatch rates can be as high as 14 pounds per pound of the targeted fish landed [9]. A high bycatch rate can cause unnecessary damage to the ecosystem, is very costly, is a waste to marine life, and is thus a high priority by fishermen to decrease it as much as possible.

    During a short fishing season, fishermen would race to catch as much as possible, and this caused large rates of bycatch. Introducing catch shares allowed the fishermen to be more efficient and developed ways to reduce the bycatch. This included using certain types of gear, circle hooks to protect sea turtles, and certain net mesh sizes to avoid young fish.

4. *Limiting Fishing Impacts on Habitats*   Before any limitations were placed on fishermen, they employed every possible technique to

maximize their catch. This would include gear and netting that would sweep the ocean floor in order to catch as much of the target fish as possible. Not only would this system include large volumes of bycatch, but it would alter the ocean ecosystem and reduce biological diversity. Sometimes this gear would get tangled, and the fishermen would simply cut it loose rather than try to recover it. This decision is made based on the revenue that could be enjoyed during the time it would take to untangle the gear.

With the introduction of catch shares and thus a better managed system, fishermen were able to change their equipment. This allowed them to meet their quota without impacting the habitat. Fishermen were able to use fewer hooks since their goal was to catch fewer fish, and without the need to race to catch fish, there was less of a chance to lose large fishing gear.

5. *Making Fishing Safer*   Without a catch limitation, fishermen would compete to maximize their catch. In many situations, they had to choose between their safety and making a living. In the United States, the occupational fatality rate among fishermen is as much as 35 times higher than all-industry averages [10]. Racing to maximize their catch, boats may fish during dangerous weather conditions and far from any potential assistance if it is needed. Quite often they would establish a grueling schedule and operate most of the day and night.

The introduction of catch shares has changed the operating methods so that fishermen no longer need to take unnecessary chances while competing with other fishermen. If a storm is forecasted, the fisherman no longer has to operate during the storm but wait for better weather, knowing that he or she has catch shares to fulfill.

6. *Improving Economic Performance*   Without the implementation of catch shares, the fishing industry could not operate efficiently. As indicated earlier, fishermen competed fiercely, operated long days, had large bycatch rates, and discarded tangled gear. These problems contributed to low revenues and high costs. "In the five years leading up to LAPPs, a study show that revenue per boat decreased, by an average of 10%. In the five years after LAPPs, revenues increased by an average of 80%" [7].

The introduction of catch shares provided many economic benefits:

- Reduced capital costs by reducing fleet sizes
- Matched fishing resources with total allowable catch levels
- Increase in boat yields

- Better planning of season
- Extended seasons lead to more consistent landing of fish
- Processors able to keep fish fresh
- Produce higher-quality seafood
- Increase in the value of catch shares
- Sustainable fisheries

NEW SUSTAINABLE FISHING POLICY FOR EUROPE

While Europe is not usually considered having a large fishing industry, there is still sufficient interest in sustainability that a new Common Fisheries Policy was introduced in early 2013. The policy being implemented would outline the future for the European seas through 2020. Basically, the question is how can sustainable fishing be ensured in Europe? [11]

Marine products are seen as a healthy component of human nutrition. They contribute fully to the diversity of the food on offer in Europe. So this policy is not about a mundane economic activity, but a vital one.

Two proposals, recommended as being protective of resources and focused on the objective of maximum sustainable yield, were examined as part of this reform: transferable fishing concessions (TFCs) and "zero discards." TFCs are the basic application of the economic precept that a market strikes an optimum balance through the free allocation of resources among its individual players. It comes down to telling fishermen that together they are managing the resources badly and therefore suggesting that they assume individual responsibility for managing a quota. This proposal harbors the seeds of the decline of the fishing industry as nothing would prevent players other than fishermen from acquiring quotas. And nothing would force them to use them to fish. In view of the conflicts about use that arise from sharing the sea, this will ultimately empty the sea of its fishermen so that the sea can be exploited for other purposes, such as energy or oil. Is that sustainable fishing?

As for zero discards, this is archetypical of a bogus good idea. By making it compulsory to land unwanted catches, being proposed is the establishment of an animal meal industry for fish farming. It fails to take account of the loss that these non-discards would mean for the marine ecosystem. Even when they are dead, discarded fish partly feed the ecosystem on which the species in demand thrive. Moreover, apart from this loss to the natural food chain, innovations in the selectivity of fishing gear will allow for the continually reduction of bycatches. The animal meal markets created in this way would therefore not be viable: they would have to serve other markets and therefore prompt an expansion of industrial fishing, itself indecent given that it involves catching fish to be ground into meal in order to feed farmed stocks, whereby five tons are caught for every one ton of fish farmed. If the goal is to replace fisheries with fish farming, this could be a detriment of marine ecosystems. This could be a consequential reality of zero discards.

Sustainable fishing means renewing fish stocks, fishing vessels, and fishermen. These new proposals would mean making fishermen responsible for managing resources, with an awareness of this food security challenge. The intent was to promote innovation in the areas of fish-stock monitoring, fishing vessels that consume less fuel, and the selectivity of fishing gear and techniques, thus improving the economic performance of firms and reducing the sector's environmental footprint.

The introduction of the Common Fisheries Policy was an opportunity to set Europe permanently on the path towards healthy eating, while respecting the environment and securing its future. If this policy were to be enforced, a sustainable fisheries industry would be possible in Europe.

CASE STUDY: FISHBANKS LIMITED

The FishBanks Ltd. game was developed by Dennis Meadows, University of New Hampshire, after many years of testing and development.

This game is about sustainable resource management. It uses as its basis marine fisheries to represent renewable resources, and lessons can be extrapolated to others such as forestry and soil. FishBanks [12] is based on a computer model developed through application of a systems analysis technique called system dynamics (SD). SD is a comprehensive approach to the representation, diagnosis, and change of behavior patterns in complex, dynamic systems. The SD method is based on concepts of information feedback, and it employs computer simulation of feedback models representing real-world issues.

In this game, players take on the role of fishing companies operating in shared coastal and deep-sea fisheries. Their goal is to achieve the greatest possible assets by the end of the game. The assets equal the sum of the accumulated bank balance plus the salvage value of the ships at the end of the final year in the game. How they go about it leads to rich learning around both social and environmental aspects of sustainability.

The resources at the beginning of the game consist of a fleet of ships, a bank account, and access to two offshore fishing areas.

In each round, the team must determine its fleet size by deciding whether to bid for ships at auction, make ship trades with other teams, order new ships to be constructed by the shipyard, or maintain the current size of the fleet. Then each team must decide how to allocate the ships among the coastal and deep-sea fishing areas and the harbor.

The bank balance is increased by income from fish and ship sales and decreased by expenditures for ship purchases and operations. Additionally, the account is subject to interest earning and charges. The total assets equal the sum of the bank balance plus the salvage value of the ships at the end of the game.

Income can be earned by selling fish catch at a fixed price per fish, selling ships to other teams at a negotiated price, and earning interest on the minimum bank balance during the years it remains positive.

Expenses are incurred by buying ships at auction, buying ships from other teams at a negotiated price, or ordering the construction of a new ship. In addition, there are operating expenses that vary depending on which fishing area the ship is operating.

Each team must decide in which fishing area to operate. The biological maximum for each area is different but known to the fishermen. With little overharvesting, each area operates at its maximum population. But as the teams compete for fish, as the population decreases, the productivity decreases and the population decreases—possibly below a level where its maximum can no longer be achieved. Data are also provided to show the effectiveness of each ship in each fishing area as a function of fish population.

This computer-simulated game can be obtained from various sources.

REFERENCES

1. Hinrichsen D. *Coastal Waters of the World: Trends, Threats, and Strategies*. Washington, DC: Island Press; 1998.

2. Wilson, EO. *Biophilia*. Cambridge, MA: Harvard University Press; 1984.

3. Available at http://www.economicsonline.co.uk/Market_failures/Depletion_of_fish_stocks.html. Accessed 2013 Jul 2.

4. Handwerk B. U.S. chefs join campaign to save Chilean sea bass, *National Geographic News*, May 22, 2002.

5. John G. Troubled waters: a special report of the seas, *The Economist*, January 3, 2009.

6. International Pacific Halibut Commission Annual Report 1990. Available at http://www.iphc.int/publications/annual/ar1990.pdf. Accessed 2013 Jul 2.

7. Sustaining America's fisheries and fishing communities: an evaluation of incentive-based management, environmental defense. Available at http://www.edf.org/sites/default/files/sustaining-fisheries.pdf. Accessed 2013 Jul 2.

8. America's living oceans: charting a course for sea change: a report to the nation, Pew Oceans Commissions, Arlington, VA; 2003.

9. Alverson DL, Freeber MK, Murawski SA, Pope JG. A global assessment of fisheries bycatch and discards, FAO fisheries technical paper no. 339, Food and Agriculture Organization of the United Nations, Rome.

10. Sitka AK. Proceedings of the Second International Fishing Industry Safety and Health Conference; 2003 Sept 22–24. Available at http://www.cdc.gov/niosh/docs/2006–114/. Accessed 2013 Jul 2.

11. Karleskind P. Sustainable fishing is possible in Europe, *Europolitics*, February 4, 2013.

12. Available at http://fishbanksgame.blogspot.com/2011/07/how-to-order-your-fishbanks-game.html. Accessed 2013 Jul 2.

CHAPTER 11

Environmentally Effective Buildings

In 1981, W. McDonough [1] started his architectural practice, and his first major commission was the design of the Environmental Defense Fund (EDF) headquarters in New York City in 1984 [2]. Being very concerned with the environment, EDF made a stipulation that the design of the building should provide the highest possible air quality for the benefit of the EDF employees. He was able to accomplish his goal while working with David Gottfried, a construction manager, and Michael Italiano, an environmental attorney.

During the next few years, David Gottfried and Michael Italiano continued their interest in environmentally efficient and effective buildings. They conducted many informal meetings with Rocky Mountain Institute, Green America, Carrier Corporation, Herman Miller, and Interface. In 1993, Gottfried, Italiano, and S. Richard "Rick" Fedrizzi, the marketing manager for Carrier's parent company, United Technologies Corporation, founded the US Green Building Council (USGBC) with Fedrizzi as the chair of the USGBC.

As the organization was created for the development of environmentally effective, as opposed to efficient per *Cradle to Cradle* [1], buildings, the first task for the USGBC was to define such a "green" building. In simple terms, a green building was defined by five factors:

- Energy efficiency—the consumption of energy for the construction of the building and the operation of the building must be at a minimum. Use of renewable energy for any portion of the building's operation enhances its definition.
- Water efficiency—the consumption of water for the construction of the building but more important for its operation should be at a minimum.

Practical Sustainability Strategies: How to Gain a Competitive Advantage,
First Edition. Nikos Avlonas and George P. Nassos.
© 2014 John Wiley & Sons, Inc. Published 2014 by John Wiley & Sons, Inc.

- Efficiency in materials—there should be efficiency in the use of materials in the construction of the building. This includes the use of recycled materials, sourcing materials from nearby locations, and the use of renewable materials.
- Reducing health impacts—the design and construction of the building should minimize the impact on human health by incorporating the highest quality of air and minimizing the emission of toxic vapors from the materials or coatings.
- Reducing building impact on the environment—considerations in the design of the building should include minimizing the impact on the environment in terms of air, water, soil, and vegetation.

Once the organization defined a green building, its next initiative was the creation of the Leadership in Energy and Environmental Design (LEED), a certification program for new buildings that would include a more specific definition of a green building. USGBC had six primary goals [3] in mind when it developed LEED:

1. Define "green building" by establishing a common standard of measurement.
2. Promote integrated, whole-building design practices.
3. Recognize environmental leadership in the building industry.
4. Stimulate green competition.
5. Raise consumer awareness of green building benefits.
6. Transform the building market.

The first version, called LEED 1.0, was launched in 1998 and provided new buildings for earning up to 40 points. Depending on the number of points awarded to a new building, it could earn a bronze, silver, or gold award. However, there were many shortcomings, and during the next two years, this initial version was subject to testing and review, and it led to the development of LEED 2.0, which expanded to 69 points. Although these standards were designed primarily for new construction, they were also applied to existing construction in the event there was a need to convert an existing building. More testing and reviews resulted in LEED 2.1 in early 2003.

It wasn't until November 2005 when USGBC introduced LEED-NC 2.2, which was specific for new construction. Modifying an existing building to achieve LEED status is considerably different than constructing a new building, so a new set of standards was developed. For instance, points might be awarded for the material of construction of a new building, but this would not be possible for an existing one.

In addition to LEED for new construction, USGBC also developed the following standards:

- LEED-EB for existing buildings
- LEED-CI for commercial interiors
- LEED-CS for core and shell
- LEED-H for houses
- LEED-ND for neighborhood development

USGBC also developed a series of application guides for the different types of buildings and building complexes. Some of the application guides are for:

- Retail
- Multiple buildings
- Campuses
- Lodging
- Healthcare
- Laboratories
- Schools

The maximum number of points that can be earned by new construction through LEED-NC 2.2 was set as follows:

- Sustainable sites   14
- Water efficiency   5
- Energy and atmosphere   17
- Materials and resources   13
- Indoor environmental quality   15
- Innovation and design process   5

Total   69

Some of the approaches that can lead to the points are:

- Sustainable sites—not on prime farmland; not lower than five feet above the 100-year floodplain; not within 100 feet of wetlands or brownfield; bicycle storage and shower for riders; preferred parking for carpool vans, hybrid or electric automobiles; green roof.
- Water efficiency—use captured rainwater; no irrigation; treat water on-site.
- Energy and atmosphere—no use of chlorinated fluorocarbons; use renewable energy.

- Materials and resources—recycle waste materials; divert demolition waste from landfill; source materials from within 500 miles of site.
- Indoor environmental quality—use low VOC coatings, recycled carpeting, and operable windows and daylight use.
- Innovation and design process—retain a LEED accredited professional (AP); innovations are not included in LEED.

By earning as many points as possible, a newly constructed building can receive an award as follows:

- Certified   28–32 (under LEED 2.0, this level was called bronze)
- Silver   33–38
- Gold   39–51
- Platinum   52–69

For an existing building, the total number of points available for earning a LEED-EB award is significantly more than that for new construction:

- Sustainable sites   14
- Water efficiency   5
- Energy and atmosphere   23
- Materials and resources   16
- Indoor environmental quality   22
- Innovation and design process   5

Total   85

You will note that the larger number of points for the LEED-EB is confined to just three of the six categories, energy and atmosphere, materials and resources, and indoor environmental quality. It is in these categories that challenges are faced to convert an existing building to a LEED-certified building.

Despite several upgrades of the LEED system over a seven-year period, there were still several shortcomings:

- Go after easy points—since there are alternative means to earn points, consultants might recommend, for example, installing a bicycle rack and a shower rather than installing a green roof. Either option is worth one point, but the costs are much different.
- Regional differences—there is a big difference between, say, Arizona and Illinois in terms of solar energy or water availability.
- Need weighting factors—while some enhancements earn the same number of points, the benefits may differ. Bicycle racks and green roofs earn the same points, but green roofs are more beneficial in terms of water retention and building cooling.

- LCA is missing—there is no life cycle assessment made on the various alternatives.
- Definition of green material or product—without standards for green products, there is no consistency for awarding points for the products used.
- Analysis of cost versus benefit—it would be helpful to determine the environmental, social, and/or economic benefit as a function of the cost.
- Point totals confusing—there is no consistency in the total points as a function of the type of building.

Many of the shortcomings and other improvements were incorporated in LEED 3.0, which expanded the certification requirements to seven categories allowing for a total of 110 points:

- Sustainable sites   26
- Water efficiency   10
- Energy and atmosphere   35
- Materials and resources   14
- Indoor environmental quality 15
- Innovation in design   6
- Regional priority   4

Total   110

Regional priority is a new category that was added in this version. This new category takes into consideration the environmental priorities of different regions within the United States.

Within each of the seven categories are two sublevels of requirements. The first sublevel of requirement is the "prerequisite," which consists of one to three prerequisites for each of the first five categories with total of eight in all. The second level of requirements is the credits, which can be earned only if the prerequisites are achieved.

In addition to increasing the total number of points that can be earned to 110, LEED 3.0 was also modified to provide for a consistent number of points for each certification level regardless of the building category. Based on LEED 3.0, the required points are now:

- Certified   40–49
- Silver   50–59
- Gold   60–79
- Platinum   80+

The next version of LEED was released in 2012 and named LEED 4.0 [4].

Earning any level of accreditation is an excellent achievement of creating an efficient and healthy building. Many of the building owners seek LEED certification not only for the benefits but also for marketing purposes. If it is a rental building, the owner can attract quality tenants and can charge higher rates. Other owners seek the accreditation for the human and social benefits. Schools with daylighting showed improved test scores, while stores with skylights can achieve greater sales. Productivity in business settings is also increased with improved lighting and clean air circulation. Reduced absenteeism due to illnesses not only improves productivity, but it can also reduce health insurance costs.

An example of a LEED Platinum building is the Bank of America Tower with the following features:

1. Water from sky and earth—rainwater is collected and used for flushing toilets, irrigating the green roof, and running the air conditioning.
2. Daylight savings—floor-to-ceiling windows reduce the lighting costs and improve productivity.
3. Chill factor—chilled air is pumped into a space below raised floors, and as it warms up, it rises and pulls more cool air with it.
4. Aired out—air is brought into the building from above the tenth floor as it is free of tailpipe emissions from vehicles. Despite the use of cleaner air, it is still filtered to remove particulate matter, allergens, ozone, and other compounds.
5. Homemade juice—the Bank of America Tower contains its own power plant running on natural gas. The efficiency is relatively high as there are no losses from shipping electricity over power lines, and the waste heat is reused.
6. No parking—there is very limited parking for automobiles, but rather the lower level is accessible to a network of public transportation. There is also secured bicycle parking.
7. Ice storage—during the night when electricity costs are much less, ice is produced to provide cooling for the air-conditioning system.

LEED PROJECT CERTIFICATION PROCESS

The owner of a building or project must determine which particular LEED rating system will be the most appropriate for the proposed project (3, p. 24). Once the rating system has been selected, there is a five-step process required to achieve certification:

1. Project registration—this is the first step, and it must be done through the Green Building Certification Institute (GBCI), which has assumed the administrative responsibility for all LEED project certifications. The registration process itself, which is managed online, consists of seven steps:

    (a) Eligibility—verifying that the project meets all the requirements.

    (b) Rating system selection—actual selection of the rating system to be used.

    (c) Rating system results—project administrator receives an online LEED scorecard that identifies all the prerequisites and possible credits.

    (d) Project information—project administrator inputs information on the project.

    (e) Review—computer reviews the project information and registration information for completeness.

    (f) Payment—administrator pays necessary fees.

    (g) Confirmation—computer issues a final confirmation that the project is registered.

2. Application preparation—this step is conducted online over a period of time. As each activity for a prerequisite or credit is completed, the information requested is entered online.

3. Application submittal—the application submittal step requires two types of information for review by the GBCI. One requires information on the design of the building including site location and building design information. The second type of information shows how the project meets with compliance during the actual construction process.

4. Application review—this is the review by the GBCI to determine whether the project actually meets the requirements. The GBCI will reward the points or deny them. If there are any points denied, the project administrator can appeal the decision, accept the decision, or add additional points in other categories to make up the deficit.

5. Certification award—the final step is the actual award of certification by the GCBI.

LEED ACCREDITED PROFESSIONAL

A LEED AP is highly recommended for anyone pursuing a project seeking LEED certification. This professional has been trained to understand fully the certification process and what is needed to achieve certification. From 2001 to

2009, exams were given by the USGBC to those interested in becoming a LEED AP, and each year the exam was more intensive as the project requirements became more specific as the process went from LEED 1.0 to LEED 2.2.

In 2009, the GBCI took over the administration of this process and developed a new pathway for becoming an AP [5]. The first step is to become a Green Associate, which is intended for those professionals who want to demonstrate green building expertise in nontechnical fields of practice. There is no need, however, to start with the Green Associate credentials as one can pursue the LEED AP accreditation directly. While LEED AP was the only position possible through 2009, now it is possible to earn LEED AP for specific types of projects. Available today are the following:

- LEED AP BD+C—for building design and construction and includes new construction, schools, and core and shell
- LEED AP ID+C—for interior design and construction and includes commercial interiors
- LEED AP homes—for the design and construction of high-performance green homes
- LEED AP O+M—for the operation and maintenance of existing buildings
- LEED AP ND—for the design and development of neighborhoods that meet the high levels of environmentally responsible and sustainable development

Additional AP positions will, no doubt, be developed as the LEED process continues to evolve.

LIVING BUILDING CHALLENGE

In 2007, a Seattle architect, Jason McLennan, introduced the Living Building Challenge (LBC) whereby new structures would produce all of their own energy and use only water that falls on-site. This new building standard has been adopted by the Cascadia Region Green Building Council for Oregon, Washington, and British Columbia. Builders would be required to use sustainably sourced materials and avoid toxic materials like asbestos, mercury, and polyvinyl chloride (PVC). They must build on previously developed sites and meet measurements of livability, social equity, and beauty.

When Cascadia released the LBC, its objective was not to compete with LEED. In fact, at that time the thought was to provide some ideas for the development of LEED 3.0. The plan was to make the LBC requirements more sustainable than that for LEED Platinum. The LBC is not a rating system, so there are no points to be earned. The basis of LBC was a series of 16 prerequisites of which all must be met. It has since been increased to 20 prerequisites through version LBC 2.1 [6]:

1. Limits to growth—projects may only be built on previously developed sites, either greyfield or brownfield, and may not be built within 50 feet of wetlands or adjacent to sensitive ecological habitats.
2. Urban agriculture—the project must integrate opportunities for agriculture appropriate to its scale and density.
3. Habitat exchange—for each acre of development, an equal amount of land must be set aside as part of a habitat exchange.
4. Carefree living—the project must contribute towards the creation of walkable, pedestrian-oriented communities.
5. Net zero water—100% of the occupants' water use must come from captured precipitation or reused water that is appropriately purified without the use of chemicals.
6. Ecological water flow—100% of storm water and building water discharge must be handled on-site.
7. Net zero energy—100% of the building's energy needs to be supplied by on-site renewable energy on a net annual basis.
8. Civilized work environment—every occupiable space must have operable windows that provide access to fresh air and daylight.
9. Healthy air—all buildings must meet specific criteria for air sources and control depending on the space itself.
10. Biophilia—the project must be designed to include elements that nurture the innate human attraction to natural systems and processes.
11. Materials red list—the project cannot contain certain materials or chemicals such as lead, cadmium, mercury, PVC, or chlorofluorocarbons (CFCs).
12. Embodied carbon footprint—the embodied carbon footprint of its construction must be offset through carbon credits.
13. Responsible industry—all wood must be Forest Stewardship Council (FSC) certified or from salvaged sources.
14. Appropriate sourcing—depending on the material or service, it must be obtained within a certain radius that can vary from 250 miles (heavy materials) to 7000 miles (renewable energy technologies).
15. Conservation and reuse—construction waste must be diverted from landfills that can vary, depending on the waste, from a minimum of 80% up to 100%.

16. Human scale and humane places—the project must be designed to create human-scaled rather than automobile-scaled places, so that the experience brings out the best in humanity and promotes culture and interaction.

17. Democracy and social justice—all primary transportation, roads, and nonbuilding infrastructure that are considered externally focused must be equally accessible to all members of the public.

18. Rights to nature—the project may not block access to, nor diminish the quality of fresh air, sunlight, and natural waterways for any member of society or adjacent developments.

19. Beauty and spirit—the project must contain design features intended solely for human delight and the celebration of culture, spirit, and place appropriate to the function of the building.

20. Inspiration and education—educational materials about the performance and operation of the building must be made available to the public.

The philosophy of the LBC is for every single act of design and construction to make the world a better place. This should be the philosophy of all designers and builders.

WORLD'S GREENEST BUILDING

On Earth Day on April 22, 2013, the Bullitt Foundation opened its new building, the Bullitt Center (Fig. 11.1), in Seattle, Washington. The 50,000 square-foot building was considered by the builders to be the greenest building in the world as they bypassed the USGBC's LEED certification in favor of the more stringent 20 prerequisites of the LBC.

The Bullitt Center's approach towards environmental sustainability started with the design of the site. Cisterns store rainwater, and "grey water" from sinks and showers funnel through the building's green roof. The Bullitt Center is flanked by a planting strip that makes the approaching sidewalk more pleasant for local workers and residents. Solar arrays provide as much electricity as the building requires. Medium-height sidewalk plantings also create a physical separation between pedestrians and vehicle traffic. The building's planners chose a thriving residential neighborhood intermixed with shops, restaurants, and bars for the opportunity to add more commercial space. This approach enhances the vision of people living closer to their place of work so that commutes are very short or, better yet, sufficiently close to walk to the office.

In tune with the ideals behind the LBC, the Bullitt Center took inspiration from nature and created a work environment that is practical yet also healthy for its inhabitants. Architectural details that are aesthetically pleasing yet

FIGURE 11.1   The New *Bullitt Center*.

FIGURE 11.2   Solar Energy for the *Bullitt Center*.

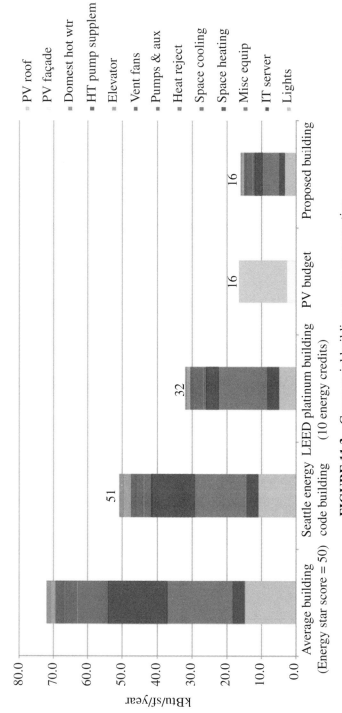

FIGURE 11.3 Commercial building energy consumption.

practical include higher ceilings (eliminating an additional floor possible under local building codes) and a central glass-enclosed staircase that encourages tenants to use the stairs instead of the elevator. Exposed wood, FSC certified, is a reflection of the local Pacific Northwest natural environment.

In order to meet the water requirements of the LBC, the building must be designed to provide its own supply of water through rainwater capture. Until that really happens, water will be supplied by the City of Seattle. It is not only a question of quantity, but the building owners must also prove that the water is safe for health reasons.

The Bullitt Center also needs to generate as much energy as it uses to comply with the LBC, so it is topped off with a giant solar array that is bigger than the building's footprint (Fig. 11.2). It is tied into the grid but is projected to be electricity neutral, using a third of the energy of normal buildings. The building is heated using ground source heat pumps. As a result of these energy efficiency features, the Bullitt Center consumes 75% less energy than the average commercial building as shown in Figure 11.3.

Some of the other features include great windows that open, high ceilings for lots of natural light, greener concrete, careful choice of materials, and bike racks instead of parking spaces. The elevators were purposely designed to move slowly in order to encourage the use of the staircases, which are next to windows with beautiful views. One of the strictest LBC requirements is that of durability. This building was designed to last 250 years. No one today can be sure that it will meet that requirement. In fact, the building cannot be LBC certified until it has been occupied and operating for a minimum of one year.

The vision of the Bullitt Foundation was to develop a model to meet the goal of the LBC where every feature and benefit of the building design will make the world a better place. If all new buildings in the world met this goal, the earth would certainly be a better place.

REFERENCES

1. McDonough W, Brangart M. *Cradle to Cradle*. New York: North Point Press, 2002.
2. Available at http://en.wikipedia.org/wiki/William_McDonough. Accessed 2013 Jul 2.
3. Farmer G. *Contractor's Guide to LEED-Certified Construction*. Clifton Park, NY: Delmar Cengage Learning; 2012.
4. Available at http://www.usgbc.org/credentials. Accessed 2013 Jul 2.
5. Available at http://www.usgbc.org/DisplayPage.aspx?CMSPageID=1815. Accessed 2013 Jul 2.
6. Available at https://ilbi.org/lbc/LBC%20Documents/lbc-2.1. Accessed 2013 Jul 2.

CHAPTER 12

Green Chemistry, Nanotechnology, and "Big Hairy Audacious Goal"

In the preceding chapters, we have provided a summary of various strategies that can be employed by corporations, government agencies, and nonprofits to offer products and services of the highest quality without having a negative impact on the environment. In the next decades, it will be extremely critical to be aware of the declining health of the environment and that everything we do should improve its health before it is too late. In addition to employing strategies like those described in the earlier chapters, there may be many others that can be integrated into an organization's operation to accomplish the same goal. Here we summarize three additional strategies.

GREEN CHEMISTRY

Green chemistry is a revolutionary approach to the way that products are made; it is a science that aims to reduce or eliminate the use and/or generation of hazardous substances in the design phase of material development. It requires an inventive and interdisciplinary view of material and product design. Green chemistry follows the principle that it is better to consider waste prevention options during the design and development phase than to dispose or treat waste after a process or material has been developed. This does not reduce the amount of waste material generated, but it reduces the quantity of feedstock required to manufacture the product.

Practical Sustainability Strategies: How to Gain a Competitive Advantage,
First Edition. Nikos Avlonas and George P. Nassos.
© 2014 John Wiley & Sons, Inc. Published 2014 by John Wiley & Sons, Inc.

For a technology to be considered green chemistry, it must accomplish three things:

- It must be more environmentally benign than existing alternatives.
- It must be more economically viable than existing alternatives.
- It must be functionally equivalent to or outperform existing alternatives.

Green chemistry presents industries with incredible opportunity for growth and competitive advantage. This is because there is currently a significant shortage of green technologies: it is estimated that only 10% of current technologies are environmentally benign; another 25% could be made benign relatively easily [1]. The remaining 65% have yet to be invented! Green chemistry also creates cost savings: when hazardous materials are removed from materials and processes, all hazard-related costs are also removed, such as those associated with handling, transportation, disposal, and compliance.

Through green chemistry, environmentally benign alternatives to current materials and technologies can be systematically introduced across all types of manufacturing to promote a more environmentally and economically sustainable future.

In 2007, John Warner and Jim Babcock partnered to found the first company completely dedicated to developing green chemistry technologies, the Warner Babcock Institute for Green Chemistry. The institute was created with the mission to develop nontoxic, environmentally benign, and sustainable technological solutions for society.

The institute works with corporations to find a more sustainable process for the manufacture of chemical products. The methodology by which the institute develops these processes is based on 12 principles:

1. Pollution prevention—It is better to prevent waste than to treat and clean up waste after it is formed.
2. Atom economy—Synthetic methods should be designed to maximize the incorporation of all materials used in the process into the final product.
3. Less hazardous synthesis—Whenever practicable, synthetic methodologies should be designed to use and generate substances that possess little or no toxicity to human health and the environment.
4. Design safer chemicals—Chemical products should be designed to preserve efficacy of the function while reducing toxicity.
5. Safer solvents and auxiliaries—The use of auxiliary substances (solvents, separations agents, etc.) should be made unnecessary whenever possible and, when used, innocuous.

6. Design for energy efficiency—Energy requirements should be recognized for their environmental and economic impacts and should be minimized. Synthetic methods should be conducted to ambient temperature and pressure.

7. Use of renewable feedstocks—A raw material or feedstock should be renewable rather than depleting whenever technically and economically practical.

8. Reduce derivatives—Unnecessary production of derivatives (blocking group, protection/deprotection, and temporary modification of physical/chemical processes) should be avoided whenever possible.

9. Catalysis—Catalytic reagents (as selective as possible) are superior to stoichiometric reagents.

10. Design for degradation—Chemical products should be designed so that at the end of their function they do not persist in the environment and instead break down into innocuous degradation products.

11. Real-time analysis for pollution prevention—Analytical methodologies need to be further developed to allow for real-time in-process monitoring and control prior to the formation of hazardous substances.

12. Inherently safer chemistry for accident prevention—Substance and the form of a substance used in a chemical process should be chosen so as to minimize the potential for chemical accidents, including releases, explosions, and fires.

When working in the various industry sectors, the approach to green chemistry may vary slightly. For example, here are some of the approaches taken for the following industry sectors:

• Energy, natural resources, and environment—Novel photon to chemical energy mechanisms may be used for the generation or storage of energy. Alternative fuels may be developed from renewable feedstocks and intermediates. Diverse approaches may be used for the sustainable reclamation and/or purification of water.

• Industrial chemicals and materials—When attempting to manufacture a specific product, there may be different approaches to achieve the ultimate chemical product. Here is where one must look at alternative feedstocks that are more sustainable and economic.

• Industrial products—It is important for the products to be void of volatile organic compounds (VOCs) as well as other chemical coatings that are prohibited by new regulations.

• Pharmaceuticals and biotechnology—New methods are sought in order to reduce the non-active compounds in a particular product while developing new synthetic methods for the active ingredient.

- Personal care and cosmetics—Must develop new materials and application mechanisms that avoid VOCs and other potentially harmful materials to the skin, hair, and nail products.
- Retail, consumer, and supply chain—Must assess the entire supply chain to be sure the materials used in the manufacture of products are sustainable and meet current and potentially future regulations.

An example of applying green chemistry to a pharmaceutical product is the synthesis of ibuprofen [2]. Ibuprofen is the active ingredient in many analgesic and inflammatory drugs such as Advil, Motrin, and Medipren. Beginning in the 1960s, ibuprofen was produced by a six-step synthesis with an atom economy of only 40%. This meant that less than half (40%) of the weight of all the atoms of the reactants was incorporated in the ibuprofen and 60% was wasted in the formation of unwanted by-products. The annual production of approximately 30 million pounds of ibuprofen by this method resulted in over 40 million pounds of waste. But during the 1990s, the BHC Company developed a new synthesis of ibuprofen with an atom economy of 77–99%. This synthesis not only produces much less waste; it is also only a three-step process. A pharmaceutical company can thus produce more ibuprofen in less time and with less energy, which results in increased profits.

In another example, the shoe company Nike identified five chemicals in its original shoe rubber that were hazardous and worked diligently to eliminate these chemicals. [3] In the original rubber, those five toxic chemicals made up 12% of the product by weight. The "green rubber" that Nike eventually created by applying green chemistry has only one of the five chemicals in it, and that chemical makes up only 1% of the product by weight. That reduced the toxic composition by 96%, or 3000 metric tons per year.

The company was also successful in reducing zinc in its shoes. Using zinc meant emitting 340 grams of VOCs for every pair of shoes during the manufacturing process. But Nike engineers discovered the zinc wasn't really that essential to the shoes. They were able to remove 80–90% of the zinc in the shoe manufacturing process—reducing toxic emissions from 340 grams per pair in 1995 to 15 grams in 2006. This is a huge difference that would never be noticed by the consumer, but would be noticed by the workers in the factories.

NANOTECHNOLOGY

When K. Eric Drexler popularized the word "nanotechnology" in the 1980s [4], he was talking about building machines on the scale of molecules, a few nanometers wide—motors, robot arms, and even whole computers—far smaller than a cell. Drexler spent the next 10 years describing and analyzing these incredible devices and responding to accusations of science fiction. Meanwhile, mundane technology was developing the ability to build simple structures on a molecular scale. As nanotechnology became an accepted concept, the meaning of the word shifted to encompass the simpler kinds of nanometer-scale technology. The US National Nanotechnology Initiative was created to fund this kind of nanotech: their definition includes anything smaller than 100 nanometers with novel properties, where a nanometer is one-billionth of a meter.

To put in perspective of just how small is a nanometer, a visual comparison is helpful [5]. If a 20-nanometer object were blown up to the size of a soccer ball, a virus would be the size of a person, a red blood cell would be the size of a soccer field, a doughnut would be the size of the United Kingdom, and a chicken would be the size of Earth. In other words, a nanoparticle is really small.

Much of the work being done today that carries the name "nanotechnology" is not nanotechnology in the original meaning of the word. Nanotechnology, in its traditional sense, means building things from the bottom up, with atomic precision. This theoretical capability was envisioned as early as 1959 by the renowned physicist Richard Feynman [6] when he said:

> I want to build a billion tiny factories, models of each other, which are manufacturing simultaneously…The principles of physics, as far as I can see, do not speak against the possibility of maneuvering things atom by atom. It is not an attempt to violate any laws; it is something, in principle, that can be done; but in practice, it has not been done because we are too big (p. 36).

Based on Feynman's vision of miniature factories using nanomachines to build complex products, advanced nanotechnology (sometimes referred to as molecular manufacturing) will make use of positionally controlled mechanochemistry guided by molecular machine systems. Formulating a roadmap for development of this kind of nanotechnology is now an objective of a broadly based technology roadmap project led by Battelle (the manager of several US National Laboratories) and the Foresight Nanotech Institute.

Shortly after this envisioned molecular machinery is created, it will result in a manufacturing revolution, probably causing severe disruption. It also has serious economic, social, environmental, and military implications. Just how can this concept be applied?

One of the greatest environmental concerns on a worldwide basis is global warming. Scientists have positively concluded that this event has and continues to occur and that it is the result of human activity. Carbon dioxide emitted from numerous sources, primarily electrical power plants and automobiles, has increased the atmosphere's retention of heat generated by the sun's rays. This is the result of a thick carbon dioxide layer in the atmosphere that is expected to remain, possibly forever. However, there may also be another phenomenon caused by pollutant particles in the atmosphere that produces an opposite cooling effect called global dimming. There are separate studies measuring the amount of sunlight passing through the atmosphere and also measuring the rate of evaporation of water. Each of these two studies indicates that there has been a change in the measurements over a period of time causing global dimming. If this is truly the case, the warming effect from carbon dioxide must be greater than we thought. And if these particles are the result of power plant emissions, and if some of these power plants are replaced by renewable energy sources, the resulting reduction in particles could subsequently increase global warming. So, perhaps, the real answer to these opposing phenomena can be found if we examine them at the nanoparticle level.

Another critical environmental issue is the depletion of our natural resources. In the book entitled *Biomimicry* (see Chapter 8), Janine Benyus suggests that we take a more careful look at nature's amazing processes to become more efficient in the manufacturing of our products. An example of biomimicry is how spiders produce several different kinds of silk for various functions, such as forming webs or rappelling from drop-offs. The properties of these spider-produced silks are astounding when compared to man-made materials. Compared on an equal weight basis, some of these spider-produced silks are five times stronger than steel and five times tougher than Kevlar, the material used in bulletproof vests. At the same time, the silk can be very elastic and stretch up to 40% of its original length, something that steel wire is incapable of doing. Just imagine if someone could learn to do what the spider does, taking a renewable, soluble material and making an extremely strong water-insoluble fiber using very little energy and generating no toxic waste. By analyzing the spider's process and reproducing it, the entire fiber industry would change dramatically. This may be another role for nanotechnology.

Ms. Benyus provides other examples of natural phenomena that are extremely interesting. These include the abalone shell that is stronger than any known ceramic. If we look at the natural design of the inner coating of these shells, we may learn how to manufacture stronger materials and more sustainably. Still another example is the adhesive created by mussels and other bivalves, allowing them to attach to almost anything, and these adhesives are waterproof. By examining these natural phenomena at the nanoscale, new doors may be opened in our continuous search for a more sustainable environment.

Another example of the application of nanotechnology is the search for the perfect sunscreen. Zinc oxide would be the perfect sunscreen ingredient if the resulting product didn't look quite so silly. Thick, white, and pasty, it once was seen mostly on lifeguards, surfers, and others who needed serious sun protection. But when the sunscreens are made with nanoparticles, they turn clear—which makes them more user-friendly.

Improved sunscreens are just one of the many innovative uses of nanotechnology, which involves drastically shrinking and fundamentally changing the structure of chemical compounds. But products made with nanomaterials also raise largely unanswered safety questions—such as whether the particles that make them effective can be absorbed into the bloodstream and are toxic to living cells [7].

While the development of nanotechnology is over 30 years old, the booming nanotech industry is less than two decades old. Nanoparticles are already found in thousands of consumer products, including cosmetics, pharmaceuticals, antimicrobial infant toys, sports equipment, food packaging, and electronics. In addition to producing transparent sunscreens, nanomaterials help make light and sturdy tennis rackets, clothes that don't stain, and stink-free socks.

The particles can alter how products look or function because matter behaves differently at the nanoscale, taking on unique and mysterious chemical and physical properties. Materials made of nanoparticles may be more conductive, stronger, or more chemically reactive than those containing larger particles of the same compound.

One of the problems is that the development of applications for nanotechnology is rapidly outpacing what scientists know about safe use. The unusual properties that make nanoscale materials attractive may also pose unexpected risks to human health and the environment. "We haven't characterized these materials very well yet in terms of what the potential impacts on living organisms could be," said Kathleen Eggleson, a research scientist at the Center for Nano Science and Technology at the University of Notre Dame [8].

Scientists don't know how long nanoparticles remain in the human body or what they might do there. But research on animals has found that inhaled nanoparticles can reach all areas of the respiratory tract; because of their small size and shape, they can migrate quickly into cells and organs. The smaller particles also might pose risks to the heart and blood vessels, the central nervous system, and the immune system, according to the US Food and Drug Administration (FDA).

Though nanomaterials have been used in consumer products for more than a decade, the FDA acknowledged for the first time in April 2012 that they differ from their bulk counterparts and have potential new risks that may require testing [7]. In draft guidelines on the safety of nanomaterials in cosmetic products, the agency advised companies to consult with the FDA to find out the best way to test their products. Rather than adopting a one-size-fits-all approach, the FDA plans to assess nano-enabled products on a case-by-case basis, according to the guidelines. "There is nothing inherently good or bad about a nanomaterial," said Chad Mirkin [8], Director of the International Institute for Nanotechnology at Northwestern University, who nevertheless thinks each class of material should be considered a new form of matter and reviewed for safety.

Proponents of nanotechnology say the potential benefits reach far beyond sunscreen. Controlling matter at the atomic scale is being hailed as the next "industrial revolution" because it could help solve everything from climate change and world hunger to energy shortages and biodiversity loss. In medicine, scientists envision microscopic robots that could swim around in the bloodstream, repairing cells and diagnosing diseases. Nanotechnology also might unleash powerful new therapeutic weapons for treating many of the worse forms of cancer, cardiovascular problems, and neurodegenerative disease.

Researchers at the University of Illinois at Urbana–Champaign College of Liberal Arts and Sciences are applying nanotechnology to a number of medical projects [5]. They are developing a sensor that will give surgeons a way to collect significant amounts of information about, say, a tumor while the patient is under the knife. They are also searching for a better way to screen for prostate cancer. They are working on a way to tag the actual cells that cause cancer and determine whether and to what extent they exist in the patient. By making it easier to study membrane proteins, researchers have developed nanodiscs that allow for probing the vast number of possible pharmaceutical targets and develop more efficient drugs.

If the new nanomaterial is proved to be safe, what are the steps to manufacture it? Manufacturing at the nanoscale is known as *nanomanufacturing*. Nanomanufacturing involves scaled-up, reliable, and cost-effective manufacturing of nanoscale materials, structures, devices, and systems. It also includes research, development, and integration of top-down processes and increasingly complex bottom-up or self-assembly processes.

In more simple terms, nanomanufacturing leads to the production of improved materials and new products. As mentioned earlier, there are two basic approaches to nanomanufacturing, either *top-down* or *bottom-up*. Top-down fabrication reduces large pieces of materials all the way down to the nanoscale, like someone carving a model airplane out of a block of wood. This approach requires larger amounts of materials and can lead to waste if excess material is discarded. The bottom-up approach to nanomanufacturing creates products by building them up from atomic- and molecular-scale components, which can be time-consuming. Scientists are exploring the concept of placing certain molecular-scale components together that will spontaneously "self-assemble," from the bottom up into ordered structures.

Within the top-down and bottom-up categories of nanomanufacturing, there are a growing number of new processes that enable nanomanufacturing. Among these are:

- Chemical vapor deposition—a process in which chemicals react to produce very pure, high-performance films.
- Molecular beam epitaxy—a method for depositing highly controlled thin films.
- Atomic layer epitaxy—a process for depositing one-atom-thick layers on a surface.
- Dip pen lithography—a process in which the tip of an atomic force microscope is "dipped" into a chemical fluid and then used to "write" on a surface, like an old-fashioned ink pen onto paper.
- Nanoimprint lithography—a process for creating nanoscale features by "stamping" or "printing" them onto a surface.
- Roll-to-roll processing—a high-volume process to produce nanoscale devices on a roll of ultrathin plastic or metal.
- Self-assembly—the process in which a group of components come together to form an ordered structure without outside direction.

As new technologies allow for smaller and smaller materials, nanotechnology will become a major force in the future.

"BIG HAIRY AUDACIOUS GOAL"

In the earlier chapters, many different strategies have been discussed to provide a means for developing, extending, or maintaining a competitive advantage while being sustainable. There are, however, ideas for sustainability that don't fit any prescribed strategy. These are adopted when "thinking out of the box" or, in other words, seeking a "big hairy audacious goal" (BHAG). Three different examples are provided to show how one might develop such a project.

Washing Machines

For many years, the washing of clothes has been accomplished by adding a detergent to hot or warm water and then rinsing with cold water. This obviously is very energy intensive from the heating of the water as well as the requirement for operating the washing machine. Usually the hot water was obtained from a central hot water tank, but some machines were developed for adding cold water and heating it in the washing machine.

The soap companies eventually produced a detergent that is effective in cold water. This certainly reduces the energy requirement for washing clothes. However, the process still requires the use of a large quantity of water that is usually discharged to a sanitary sewer rather than being reused.

Applying the BHAG philosophy, how could the clothes be cleaned with minimal energy usage and no water? Perhaps the use of some nontoxic chemical in its vapor form could be used to remove dirt and grime from clothes. This chemical in pellet form could be added to, say, a container with the clothes, the pellet activated to release the vapor, and then the clothes tumbled so that the vapor makes contact with all the clothes surfaces. This, in theory, will make clothes cleaning more sustainable, and it is expected that soap manufacturers are conducting research in this area.

Toilets

In the United States almost all toilets are furnished from the same water source as our drinking water. Is it really necessary to flush liquid waste or solid waste with such clean water? If the BHAG philosophy is applied, perhaps another source of water such as grey water would be used. But where would this water be obtained for a home? Again, thinking BHAG will lead to sources of water like collected rainwater, washing machine wastewater, dishwasher wastewater, or even water going down the drain from a shower. A simple solution would be to collect the drain water from a bathroom sink and divert it to the toilet tank. This grey-water concept has been converted to a number of different designs, one of which is shown in Figure 12.1.

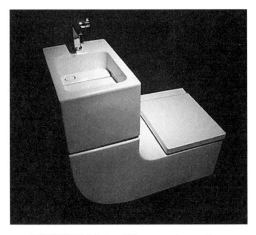

FIGURE 12.1   Water saving toilet.

Urban Farming

Prior to the nineteenth century, a typical home was four walls and a roof—a very simple structure. As people started living closer together, anyone needing more space added a second floor. But then it seemed to be more cost-effective for more than one family to live in a building with four floors and a roof. So this became the advent for apartment buildings. As land became more valuable and people continued to live in cities, developers started constructing taller buildings. So basically, starting with living at ground level, we have gone to living in high-rise buildings.

The need for more food created a demand for more agricultural land. But because the farm land is usually far from the large cities where most of the population is located, food travels great distances from its source to the dinner table estimated to be about 1500 miles. Implementing the BHAG concept, why can't agricultural land go up just like homes?

A number of entrepreneurs have begun with this concept by developing fruit and vegetable growing in previously vacant warehouses. One such grow house is UrbanPonics [9] located in the center of Chicago, Illinois, employing the science of hydroponics. An entire floor of a building can be used to grow the crops as seen in Figure 12.2, or it is possible to have multiple levels of growing trays on one floor as seen in Figure 12.3. The vegetable and/or fruit plants are grown in water only, which is a more efficient method

FIGURE 12.2   Urban farm with single layer of plants.

FIGURE 12.3   Urban farm with multiple layers of plants.

compared to water in soil where much of the water does not reach the roots of the plant. Dr. Dickson Despommier has proposed solving the food problem of the future with vertical farms and proposed one for the City of New York [10]. His proposal would allow for the use of agricultural technology like that employed by UrbanPonics but in a high-rise building of 30–40 stories, thus requiring minimum land area.

Besides producing more food for the growing population, there are numerous other advantages of vertical farming:

- Year-around crop production
- No weather problems
- All food grown organically
- Eliminates agricultural runoff
- Farming in urban cities
- Reduces fossil fuel use
- Converts black and grey water
- Provides jobs for local residents

REFERENCES

1. Available at http://www.warnerbabcock.com/green_chemistry/about_green_chemistry.asp. Accessed 2013 Jul 2.

2. Available at http://www.chemistryexplained.com/Ge-Hy/Green-Chemistry.html. Accessed 2013 Jul 2.

3. Frazier J. Nike Director of Sustainable Chemistry. Available at http://www.opb.org/news/blog/ecotrope/replacing-toxic-products-with-green-chemistry/. Accessed 2013 Jul 2.

4. Drexler KE. Molecular engineering: An approach to the development of general capabilities for molecular manipulation. Proc Natl Acad Sci 1981;78(9): 5275–5278.

5. *Alumni Magazine for the College of Liberal Arts and Sciences*, Summer 2012, College of Liberal Arts and Sciences at Illinois.

6. Feynman RP. There is *plenty* of room at the bottom. Eng Sci 1960;23(5):22–36.

7. Available at http://www.fda.gov/Food/GuidanceRegulation/GuidanceDocumentsRegulatoryInformation/IngredientsAdditivesGRASPackaging/ucm300914.htm. Accessed 2013 Jul 2.

8. Deardorff J. Scientists: nanotech-based products offer great potential but unknown risks, *Chicago Tribune*, July 10, 2012.

9. Available at http://greenurbanponics.com/index.html. Accessed 2013 Jul 2.

10. Despommier D. *The Vertical Farm: Feeding the World in the 21st Century*. New York: Thomas Dunne Books/St. Martin's Press; 2010.

CHAPTER 13

Sustainable Strategies and Beyond

There are many opportunities to implement sustainable strategies and reduce the impact on the environment, provide social benefits, and do it profitably. It is a matter of thinking on how to make a service or product more sustainable or even develop a new product or service for an unmet need.

While teaching a course titled "Business Strategy: The Sustainable Enterprise" for over 12 years, the final project for each student was to take a strategic business unit (SBU) at their place of employment and propose how to make it sustainable. The proposal must show how the new business model would possess environmental integrity, social equity, and economic viability. For those students that were not employed, they had the option to propose a project for a publicly held company or to propose a new business. Some of the proposed projects were:

- Medication delivery—proposing the reuse of bottles and tubing rather than disposal after one use.
- Farm equipment—a central logistics service providing tractor, huskers, and others to farms in a region rather than every farmer required to possess each piece.
- Eco-toys—lease toys to families as they are used for a period of time considerably less than the life of the toy.
- Endless Christmas—since both buying a plastic artificial tree from China and cutting down a real tree for two-week use are not good for the environment, why not lease a tree for the holidays and allow someone else to lease it the following year.
- Ford BOP—proposed a joint venture between Ford and Infosys, a company in India, to produce a low-cost car employing the base of the pyramid strategy.

Practical Sustainability Strategies: How to Gain a Competitive Advantage,
First Edition. Nikos Avlonas and George P. Nassos.
© 2014 John Wiley & Sons, Inc. Published 2014 by John Wiley & Sons, Inc.

- Communal refrigeration—large refrigeration and freezers to be shared by people living in small apartments in a large building.
- Highway solar panels—install solar panels along or on highways where land is not used.
- Home furnishings—expensive long-lasting home furnishings can be exchanged for a different long-lasting style.
- eClub—for those that want the latest electronic equipment, lease it and then allow someone who doesn't need the latest to buy it.
- Moo-trient—collect cattle methane waste and convert it to energy and cattle manure to fertilizer.
- Neighborhood share system—eliminates the need for all residents in a neighborhood to own maintenance equipment like lawn mowers.
- Horizontal refrigeration—migration of users to a horizontal orientation for refrigeration with present level of technology used by the chest freezers: thicker insulation and larger cooling plate.

Once the concept of sustainability is embraced, it will become automatic to think of ways to make a product or service more sustainable. Just thinking about everyday products and services, one might think of the following:

- A pizza container may be used for only five minutes, the time it takes from the restaurant to your home. Then it may not be recyclable if it is greasy, or even if it is recyclable, the process is costly for a product used for only five minutes. Why not provide returnable pizza containers?
- Most hotel rooms provide small bottles of shampoo, conditioner, and/or skin lotion. Each may be partially used and then disposed. Why not provide a large refillable container in the bathroom so the occupant uses only what is needed and provides no waste.
- When flossing teeth, a piece of floss about 12 inches long is needed for wrapping around the fingers. However, only ½ inch of the floss is actually performing the work of cleaning the teeth. The other 11½ inches is actually waste. Why not provide a device containing the floss so that only a small piece of floss is used and then advanced on the tool to expose a clean section.

Many more ideas can be developed as people accept sustainable strategies and the need for them becomes more critical. So start thinking.

PART 3

TOOLS AND METRICS

Standards and Guidelines for Managing Corporate Social Responsibility towards Sustainability

NEED FOR A SUSTAINABLE STRATEGY

The need for managing Corporate Social Responsibility (CSR) is a rising necessity nowadays, and the business world needs practical tools and guidelines to integrate sustainability into operations.

The reasons why sustainability is extremely urgent are numerous. Firstly, as pointed out by the Living Planet Report 2010 released by WWF, the ecological footprint (the measure that defines the extension of land needed to produce and dispose the goods produced) has doubled since 1966, leading humanity, in 2007, to consume the equivalent of 1.5 planets [1]. Furthermore, the gap between poor and rich, as well as problems related to health, education, and social justice, is increasing [2]. In addition to this, the situation will probably get worse, if we do not take serious measures, since the planet will be inhabited by nine billion people by 2050, according to the projection on current growth of the United Nations [3].

Sustainability management should go beyond the usual practices, and all the possible actors should take a more proactive and effective approach. Including sustainability management into the core business of companies, nongovernmental organizations (NGOs), and governments is the only way to have a more sustainable and balanced planet. If all organizations integrate sustainability into their day-to-day business, there will be good chances of facing (and maybe solving) the challenges of scarce resources and of climate change [4].

In this long and hard path towards sustainability, organizations have a fundamental key role to play. But to do so, they have to implement factors

Practical Sustainability Strategies: How to Gain a Competitive Advantage,
First Edition. Nikos Avlonas and George P. Nassos.
© 2014 John Wiley & Sons, Inc. Published 2014 by John Wiley & Sons, Inc.

that contribute to the success of a sustainable business model. The International Institute for Environment and Development has undertaken research to identify those factors that contribute to the success of business models for sustainable development. Their findings include the following:

- It is necessary that businesses start to approach different actors (companies, government agencies, and development practitioners) to build constructive and strategic alliances.
- Local communities need to be included as partners and co-designers of new models and to develop the local capacity of buying.
- The introduction of sustainable business models for development has to be self-sustaining in the long term; a significant investment of time and resources at the start is key for successful innovation and growth.
- The importance of trade-offs among different sustainable development targets (economic, social, and environmental) needs to be recognized and addressed.
- The sustainable business model has to embed an ongoing monitoring and evaluation process.

"Sustainable development is good business in itself. It creates opportunities for suppliers of 'green consumers,' developers of environmentally safer materials and processes, firms that invest in eco-efficiency, and those that engage themselves in social well-being. These enterprises will generally have a competitive advantage. They will earn their local community's goodwill and see their efforts reflected in the bottom line." Therefore, there is no reason why business (together with society) should not move towards a sustainable development [5].

MANAGING SUSTAINABILITY AND STANDARDS

In general, sustainability management is described as the intersection/interaction of the environmental, the economic, and the social sphere. Thus, sustainability is met when all these three criteria are considered and met. It is interesting to report how, in June 2010, the UN Global Compact and Accenture undertook a survey among numerous CEOs on sustainability. This survey, among other things, has underlined how most of the CEOs interviewed see sustainability as a critical issue for the future of their companies [6].

In order to create the condition to allow sustainability to flourish, there is a strong need of collaboration between different social partners in order to define a more sustainable world and lifestyle. The role of companies is indeed extremely important for this achievement; in fact, companies can significantly contribute by reducing the ecological footprint of their products as well as adopting smarter solutions for the engagement of their customers, employees, and society [7].

The "roadmap for sustainability" developed by CERES presents the main issues that a company should face to embed sustainability within its business strategy. To do so, this document identifies four areas considered key for the successful development of a sustainable business strategy, and those areas are governance, stakeholder engagement, disclosure, and performance.

Stakeholder engagement is a key aspect to include on the path towards sustainability. In fact, as it is reported in the Ceres' "roadmap for sustainability," the stakeholders are important actors in helping to identify, and reduce, the environmental and social impacts of the company. Internally, the company has to engage itself with the employees and communicate its strategy to them. The way to do it can be through the engagement of people from different business lines, areas, and regions, as well as through the creation of dedicated CSR teams. The external stakeholders are important actors, particularly for multinational companies, in order to understand the impact in other locations within the social and environmental sphere.

A comprehensive disclosure is a useful tool to disclose and manage the sustainable initiatives undertaken by the company; it is also an efficient way to include different stakeholders. In several countries, there are requirements on corporate responsibility reporting. Most of the times, the disclosures include credible and standardized metrics that deal with stakeholders concerns and express clear targets and goals.

Finally, one needs to examine the performance of the company and its relation to sustainability. According to Ceres' Report, undertaking sustainable initiatives helps the company to reduce its costs, transform wastes into resources, eliminate costs inefficiency, avoid conflict within the operations or the supply chain, and, last but not least, increase innovation [8].

There are different ways of managing sustainability, and probably the one on whose effectiveness there is more agreement is embedding sustainability into the core business of the company. Eric Lowitt wrote an interesting article in *The Guardian*, where he presented how companies such as Starbucks, UPS, Centrica, and Hitachi are adopting three common steps to embed sustainability into their competitive strategies: seek natural ways to tie sustainability and strategy together, connect sustainability to opportunity, and integrate materiality issues into competitive strategy.

The combination of sustainability and strategy design is successful when the sustainable issues faced are relevant for the company management strategy. For instance, UPS communicated with its stakeholders using a strategic framework, which was already in use and so familiar that it was integrated into other new initiatives. Secondly, linking sustainable initiatives to how they can bring new business opportunity is an important and useful process, which helps to integrate them into the corporate strategy of the company. Finally, the material assessment allows companies to prioritize the issues that are more relevant for them and their stakeholders and find a different, and more sustainable, way of solving the issue [9].

To more effectively manage sustainable practices within business, there are also numerous standards that provide help on how to integrate sustainability into the daily strategy including the Global Reporting Initiative (GRI), ISO 26000, BS 8900, and UN Global Compact.

There are also other standards and guidelines, which, even if not directly, always ask to be strongly embedded into the governance of the organization in order to be more effective and efficient in their roles of supporter of a sustainable development of the organization.

CASE STUDY ON SUSTAINABLE STRATEGY

Wal-Mart is the world's biggest retailer. After years of questionable environmental initiatives, in 2005, Wal-Mart started to build employee, nonprofit, governance agencies and develop a greener supply chain. Its goal is to decrease environmental footprint and, at the same time, increase profit. To do so, Wal-Mart intends to be supplied 100% by renewable energy, to create zero waste, and to sell products, which sustain resources and the environment. Wal-Mart realized that addressing just its operational impact would have faced just 10% of the total impact; therefore, it started to engage the external stakeholders, pushing them to join 14 "sustainable value networks," within areas of impact identified by the company "to work toward business and environmental sustainability in each area," and to create in this way the Wal-Mart's sustainable value networks.

At the core of the sustainable strategy of Wal-Mart, there is a shift towards the creation of value from long-term relationships (with nonprofits, suppliers, and other external stakeholders), which finds a profitable way to approach environmental issues, since it also increases efficiency.

Another interesting and important initiative undertaken by Wal-Mart, in order to integrate sustainability into its strategic management, is that of giving responsibility on sustainability to current employees, in order to better embed the concept within the everyday strategy and management, rather than creating a new department of sustainability. So, with the few exceptions of the small dedicated team of five persons and few people within the textile and global logistics, there is no employer of Wal-Mart that is working on sustainability full time. The reason for this is given by Tyler Elm, senior director of corporate strategy and business sustainability, which affirms that "Business sustainability isn't something you're doing in addition to your job. It is a new way of approaching your job."

> More than anything else, Wal-Mart's network approach must remain profitable if it is to be sustainable in the long run and achieve Scott's environmental goals [10].

GRI AND STAKEHOLDERS

The GRI is one of the most widely adopted methods of sustainability reporting. Global Reporting Initiative is a nonprofit organization that introduced a sustainability-reporting framework in 2000. Today, more than 3000 companies across the world use the GRI's reporting system. According to the organization's website, "The GRI Reporting Framework is intended to serve as a generally accepted framework for reporting on an organization's economic, environmental, and social performance. It is designed to be used by organizations of any size, sector, or location."

Naturally, the stakeholder approach is an essential element of this framework. The GRI Sustainability Reporting Guidelines note, "[The] multi-stakeholder approach to learning has given the Reporting Framework the widespread credibility it enjoys with a range of stakeholder groups" [11]. In other terms, it is widely accepted that businesses must consider the interests and perspectives of stakeholders in order to make significant strides in sustainability.

The GRI framework identifies "stakeholder inclusiveness" as an essential measure of sustainability. Global Reporting Initiative advises, "The reporting organization should identify its stakeholders and explain in the report how it has responded to their reasonable expectations and interests" [12]. In other terms, it is essential to measure a company's stakeholder strategy and engagement in order to properly assess and disclose its sustainability structure.

It is important to understand that stakeholder theory does not stand alone; rather, it is inextricably linked to many aspects of sustainability. For example, the concept of *transparency* is primarily fostered through stakeholder engagement. Global Reporting Initiative defines transparency as "the complete disclosure of information on the topics and indicators required to reflect impacts and enable stakeholders to make decisions, and the processes, procedures, and assumptions used to prepare those disclosures" [11]. In this sense, a company must ensure transparency in order to properly engage stakeholders and further its sustainability goals. One cannot achieve transparency or proper stakeholder engagement without the other.

GRI INTERPRETATIONS OF STAKEHOLDER ENGAGEMENT

The term "stakeholder" conjures different meanings in different contexts. Given the ambiguity of the term itself, GRI explicitly offers its interpretation of "stakeholder inclusiveness." According to GRI guidelines, "Stakeholders are defined as entities or individuals that can reasonably be expected to be significantly affected by the organization's activities, products, and/or services; and whose actions can reasonably be expected to affect the ability of the organization to successfully implement its strategies and achieve its objectives" [11]. Compared with other interpretations, GRI's definition is rather broad.

One notes that the definition itself can be interpreted in various ways, depending upon the reader's interpretation of the term "reasonably." This realization underlines the elusiveness of key terms within sustainability and alludes to the difficulties associated with establishing overarching International Standards. The ambiguity of key terms within sustainability can become problematic as different parties often subscribe to their own interpretations. In order for interested parties to collaborate and progress sustainability issues, they must agree on a single interpretation of the key terms.

THE STAKEHOLDER REPORTING PROCESS

Global Reporting Initiative's interpretation of "stakeholder inclusiveness" introduces an important challenge relating to reporting and stakeholder engagement. Considering that a sustainability report must speak to the needs and perspectives of those "reasonably" affected, how does the company determine which stakeholders it should address in its report? This question relates to the stakeholder identification.

A company must act in a systematic and organized fashion in order to respond to this challenge. Global Reporting Initiative explains, "For a report to be reliable, the process of stakeholder engagement should be documented. When stakeholder engagement processes are used for reporting purposes, they should be based on systematic or generally accepted approaches, methodologies, or principles" [11]. A company should report its process of stakeholder identification in an open and honest manner.

Similarly, a report should succinctly document how and when the company interacted with those parties and identify measures to balance divergent stakeholder expectations. Finally, it should address how stakeholder engagement affects, influences, and relates to the content of the report and the sustainability initiatives of the company [11].

GRI TESTS FOR STAKEHOLDER INCLUSIVENESS

As a comprehensive sustainability framework, GRI explicitly identifies those aspects of stakeholder engagement a company should address. The following list is an extract from GRI's guidelines, which outlines categories of information companies should disclose in order to foster proper stakeholder engagement [11].

- List of stakeholder groups engaged by the organization
- Basis for identification and selection of stakeholders with whom to engage
  - This includes the organization's process for defining its stakeholder groups and for determining the groups with which to engage and not to engage.
- Approaches to stakeholder engagement, inducing frequency of engagement by type and by stakeholder group
  - This could include surveys, focus groups, community panels, corporate advisory panels, written communication, management/union structure, and other vehicles. The organization should indicate whether any of the engagement was undertaken specifically as part of the report preparation process.
- Key topics and concerns that have been raised through stakeholder engagement and how the organization has responded to those key topics and concerns, including through its reporting

PRESENTATION OF REPORTED STAKEHOLDER DISCUSSIONS

Stakeholder engagement is an essential component of sustainability reporting. Yet, there are a variety of ways companies can choose to report their engagement efforts. While some companies opt to construct a narrative of the ways they interact with stakeholders, such as the selected case studies in this chapter, others choose a more visual or graphic approach. For example, BMO Financial Group's 2010 Sustainability Report presents its stakeholder engagement activities in outline format [12].

BMO engages regularly with a wide range of stakeholders as we work to identify key issues affecting sustainability in all of its dimensions—social, environmental, and economic. The valuable insights provided by these diverse groups guide us in better managing our business to meet their needs.

Our responsibilities to our stakeholders are as follows:

CUSTOMERS

- Satisfaction: Our vision, defining great customer experience, means we work to exceed customers' highest expectations.
- Trust: Build customer relationships based on reciprocal trust, honesty, integrity, open dialog, and mutual respect.
- Access: Provide convenient access to banking services wherever our customers live and regardless of differences in abilities.
- Products: Provide financial products and services that address our customers' needs at every stage in their lives.

COMMUNITIES

- Investment: Through donations and sponsorships, seek opportunities to support initiatives that improve the quality of life in communities where we live, work, and do business.
- Support: Maintain our commitment to supporting BMO employees' involvement as volunteers and donors in charitable campaigns and programs.

EMPLOYEES

- Talent: Cultivate an environment where all employees are encouraged to explore their potential and contribute to their fullest—and where strong performance is rewarded.
- Leadership: Develop current and future leaders' capabilities at key stages in their careers, using a consistent leadership framework across all lines of business.
- Inclusion: Promote an open and supportive workplace and insist on diversity, equity, and inclusion in every area of the organization.
- Well-being: Foster an environment that creates opportunities for employees, where change can be viewed positively and feedback is frequently sought. Advocate work–life effectiveness and other approaches that provide flexibility to employees, with managers committed to providing coaching and guidance.

SHAREHOLDERS

- Performance: Maximize returns to investors with sustainable improvements in performance achieved through effective growth strategies and sound fiscal management.
- Governance: Maintain responsible, rigorously ethical, and fully accountable corporate governance standards and practices.
- Transparency: Provide financial disclosure and environmental, social, and governance information that is clear, comprehensive, and meaningful to all shareholders.

REGULATORS

- Compliance: Remain compliant with all relevant laws and government regulations in the countries where we operate.

NGOs

- Responsibility: Address concerns raised by various NGOs by maintaining responsible lending practices and considering the environmental and social impacts of our business decisions.

SUPPLIERS

- Fairness: Strive to ensure equity in the selection and management of all vendors supplying goods and services to BMO.

ISO 26000 FRAMEWORK

Although GRI is among the most popular and highly regarded CSR frameworks, it is by no means the solitary golden standard of sustainability reporting. In 2010, the International Organization for Standardization (ISO) published ISO 26000, which introduced a new set of "Social Responsibility" guidelines. International Organization for Standardization is the world's largest developer and publisher of International Standards, as it is a network of national standards institutes from 163 countries [13]. The members of this vast network, represented by industry, government, labor, consumers, NGOs, and service representatives, spent five years creating the ISO 26000 standard [14].

According to ISO, "The International Standard ISO 26000:2010, *Guidance on Social Responsibility*, provides harmonized, globally relevant guidance for private and public sector organizations of all types based on international consensus among expert representatives of the main stakeholder groups, and so encourage the implementation of best practice in Social Responsibility worldwide" [15]. In more succinct terms, this initiative provides flexible guidance for organizations of all types on sustainability strategies.

International Organization for Standardization explicitly states that its 26000 should not be used for certification. According to the website, "Any offer to certify, or claims to be certified, to ISO 26000 would be a misrepresentation of the intent and purpose and a misuse of this International Standard" [15]. As ISO 26000 does not issue requirements, only detailed guidelines, the framework is not appropriate for certification.

The fifth clause of ISO 26000 is entitled "Recognizing Social Responsibility and engaging stakeholders." The ISO sustainability framework, like GRI, emphasizes the importance of stakeholder engagement. In fact, the ISO standard identifies stakeholder engagement as one of two "fundamental" practices of Social Responsibility. These fundamentals include "an organization's recognition of its Social Responsibility, and its identification of and engagement with its stakeholders" [16].

In addition, the framework establishes seven core subjects, the first of which is "accountability." In this context, accountability describes the company's responsibility to society as well as those affected by its decisions and operations. Evidently, this term relates directly to the stakeholder approach. The fourth core area is "respect for stakeholder interest." In conclusion, ISO "provides guidance on the relationship between an organization, its stakeholders and society, on recognizing the core subjects and issues of Social Responsibility and on an organization's sphere of influence" [17].

UNITED NATIONS GLOBAL COMPACT INITIATIVE

The United Nations Global Compact is another important initiative that serves to promote sustainability reporting. The compact is a multinational policy plan to encourage companies to act in a socially responsible manner. According to the official website, "The UN Global Compact is a strategic policy initiative for businesses that are committed to aligning their operations and strategies with 10 universally accepted principles in the areas of human rights, labor, environment and anti-corruption" [18]. The Compact's framework boasts more than 8700 corporate participants, 6000 businesses, and stakeholders from 135 countries making the UN Global Compact the world's largest voluntary CSR reporting initiative. Following are the 10 principles of UN Global Compact listed in detail:

Human Rights

- Principle 1: Businesses should support and respect the protection of internationally proclaimed human rights.
- Principle 2: Make sure that they are not complicit in human rights abuses.

Labor

- Principle 3: Businesses should uphold the freedom of association and the effective recognition of the right to collective bargaining.
- Principle 4: The elimination of all forms of forced and compulsory labor.
- Principle 5: The effective abolition of child labor.
- Principle 6: The elimination of discrimination in respect of employment and occupation.

Environment

- Principle 7: Businesses should support a precautionary approach to environmental challenges.
- Principle 8: Undertake initiatives to promote greater environmental responsibility.
- Principle 9: Encourage the development and diffusion of environmentally friendly technologies.

Anticorruption

- Principle 10: Businesses should work against corruption in all its forms, including extortion and bribery.

Overall, the Global Compact pursues two complementary objectives:

1. Mainstream the 10 principles in business activities around the world.
2. Catalyze actions in support of broader UN goals, including the Millennium Development Goals (MDGs).

With these objectives in mind, the Global Compact has shaped an initiative that provides collaborative solutions to the most fundamental challenges facing both business and society. The initiative seeks to combine the best properties of the UN, such as moral authority and convening power, with the private sector's solution-finding strengths, and the expertise and capacities of a range of key stakeholders. The Global Compact is global and local, private and public, and voluntary yet accountable.

REFERENCES

1. Human consumption. Available at http://www.footprintnetwork.org/en/index. php/GFN/page/2010_living_planet_report/. Accessed 2013 Jul 3.

2. Why do we need sustainability? Available at http://www.northlanarkshire.gov. uk/. Accessed 2013 Jul 3.

3. United Nations. Available at http://www.un.org/News/Press/docs/2007/pop952. doc.htm. Accessed 2013 Jul 3.

4. Available at http://www.guardian.co.uk/sustainable-business/blog/systemic-change-sustainability-business-strategy. Accessed 2013 Jul 3.

5. IISD Business Strategies for Sustainable Development. Available at http://www. iisd.org/business/pdf/business_strategy.pdf. Accessed 2013 Jul 3.

6. Accenture CEO's survey.Available at http://www.bizjournals.com/triad/print-edition/2011/02/18/sustainability-a-necessity-not-a.html. Accessed 2013 Jul 3.

7. Business Social Responsibility.Available at http://www.grantthornton.ca/ resources/insights/reports/IBR_2008_-_Corporate_Social_Responsibility_ Report.pdf. Accessed 2013 Jul 24.

8. Ceres' document.Roadmap for sustainability. http://www.ceres.org/resources/ reports/ceres-roadmap-to-sustainability-2010. Accessed 2013 Jul 3.

9. Why you shouldn't have a sustainability strategy. *The Guardian*. Available at http://www.guardian.co.uk/sustainable-business/blog/integrating-sustainability-business-company-strategy. Accessed 2013 Jul 3.

10. Stanford release. Available at http://www.gsb.stanford.edu/scforum/login/ documents/OIT71.pdf and http://www.valuenetworksandcollaboration.com/ images/Walmart_Value_Networks_1_.pdf. Accessed 2013 Jul 3.

11. Sustainability reporting guidelines. GRI. Available at https://www. globalreporting.org/resourcelibrary/G3-Guidelines-Incl-Technical-Protocol.pdf. Accessed 2013 Jul 3.

12. 2010 corporate social responsibility report and public accountability statement, BMO Financial Group. Available at http://www.bmo.com/cr/images/BMO_ CRPAS2010en.pdf. Accessed 2013 Jul 3. p. 6.

13. About ISO.ISO. Available at http://www.iso.org/iso/about.htm. Accessed 2013 Jul 3.

14. Social responsibility.IFAN. Available at http://www.ifan.org/ifanportal/livelink/ fetch/2000/2035/36282/394607/social_responsibility/index-sr.html. Accessed 2013 Jul 3.

15. ISO 26000:2010.ISO. Available at http://www.iso.org/iso/iso_catalogue/catalogue_tc/catalogue_detail.htm?csnumber=42546. Accessed 2013 Jul 3.

16. ISO 26000 and the definition of social responsibility, Triple pundit. Available at http://www.triplepundit.com/2011/03/iso-26000-definition-social-responsibility/. Accessed 2013 Jul 3.

17. ISO 26000—Social responsibility. ISO. Available at http://www.iso.org/iso/iso_catalogue/management_and_leadership_standards/social_responsibility/sr_discovering_iso26000.htm. Accessed 2013 Jul 3.

18. Overview of the UN Global Compact. Available at http://www.unglobalcompact.org/AboutTheGC/index.html. Accessed 2013 Jul 3.

CHAPTER 15

The Corporation and Its Stakeholders

EXAMINING THE STAKEHOLDER CONCEPT

The relationship between the organization and its stakeholders has evolved over time. During previous years, the only groups that seemed to attract an organization's attention were those related to the production and purchase of their products or services. These groups, also known as the three-legged stool, included the employees, investors, and customers. More specifically, during the precorporate period, investors, who supplied their money in order to support the business, were the only individuals to have stake in its operations and successes. These investors/owners, typically the managers of the firm, required assistance from employees to operate the business. All in all, the procedures were less complex, but so were, moreover, the expectations among the various parties involved with the organization.

As years went by, various forces triggered society towards transformation. A principal factor that greatly contributed to the gradual change of business conduct was the recognition by the public, or society, that any organization is not the sole property or interest of the founder or owner. This and many other driving forces have created the need for businesses today to be responsive to individuals and groups they formerly viewed as powerless and indifferent towards their conduct. Thus, the stakeholder concept emerged. Contemporary stakeholder focus includes an augmented number of groups, and it reveals a more sensitive approach towards them. Modern businesses, nongovernmental organizations (NGOs), and governments demonstrate the importance of stakeholder relationships and seem to grasp the vital role they play in terms of the organization's success.

Practical Sustainability Strategies: How to Gain a Competitive Advantage, First Edition. Nikos Avlonas and George P. Nassos.
© 2014 John Wiley & Sons, Inc. Published 2014 by John Wiley & Sons, Inc.

Concurrently, during the last decades, the business world appears preoccupied with an interesting debate. This debate, namely, concerns the issue of whether corporations should focus on shareholders or stakeholders. The adoption of either one of these two prominent positions actually dictates the organizational policy and strategy and moreover provides a theoretical framework upon which organizational culture is formatted. Are corporations exclusively responsible to their shareholders and therefore their only objective and responsibility is to make profit for them, or are corporations responsible for all stakeholders?

One of the most absolute advocates of the first position is undoubtedly Milton Friedman, who in 1970 wrote: "the one and only social responsibility of business is to use its resources and engage in activities designed to increase its profit." According to this theory, shareholders provide capital to corporate managers in order to achieve certain final results, usually the maximization of share values. The shareholder theory contrasts with the stakeholder theory in the sense that it focuses particularly on value creation for shareholders. Furthermore, advocates of the shareholder theory argue that the stakeholder theory does not provide a clear idea of how the corporation should address conflicting stakeholder expectations. However, advocates suggest that the principle of value maximization, of the shareholder theory, gives a clear way to conceptualize and manage trade-offs between corporate stakeholders.

In spite of this, during the 1990s, the second position of stakeholder focus was highlighted by several important conferences on stakeholder theory. "The late Max Clarkson of the University of Toronto convened two conferences in 1993 and 1994 on this topic. In 1994, Juha Näsi convened a conference on stakeholder theory in Finland" [1]. Stakeholder focus is embedded in the idea of sustainability (corporate social responsibility (CSR)). One can argue that one of the most essential contributions of sustainability is the critical shift from shareholders to stakeholders. Indeed, businesses that align sustainability with their operations are focusing on the entire spectrum of their stakeholders rather than shareholders alone.

STAKEHOLDERS: DEFINITION—PRIMARY AND SECONDARY STAKEHOLDERS

The stakeholder theory aims to systematically address the fundamental question of which groups are stakeholders and which are not. There is no actual disagreement among scholars involved in this issue, on what kind of entity is or may potentially be a stakeholder. In general, persons, groups, neighborhoods, organizations, societies, etc., are thought to be eligible stakeholders. In order to better appreciate the concept of stakeholders, one needs to first examine the idea of a stake. A stake is "an interest or a share in an undertaking," [1] as well as a claim to a title or a right to something. The idea of a stake, therefore, lies on a continuum where on one end there is the simple interest in an undertaking and on the other end a legal claim.

Proceeding to the definition of the concept, stakeholder theorists differ considerably on whether they take a broad or narrow view of the issue. For example, R. Edward Freeman, of the University of Virginia, adopting a broader perspective, defines a stakeholder as, "any group or individual who can affect or is affected by the achievement of the organization's objectives" [2]. This definition is certainly among the broadest in the literature and lies in stark contrast to Clarkson's narrower definitions of stakeholders. In fact, Clarkson creates a distinction between voluntary or involuntary risk bearers. He writes, "Voluntary stakeholders bear some form of risk as a result of having invested some form of capital, human or financial, something of value, in a firm. Involuntary stakeholders are placed at risk as a result of a firm's activities. But without the element of risk there is no stake" [3].

Between the broad and narrow are many other efforts to define stakeholders. Nevertheless, we will address the issue with respect to Freeman's definition since it includes a broader spectrum of actors. Therefore, stakeholders can be defined as categories of people that affect or can be affected by the company's decisions, policies, and operations. These parties, as well as their interests and expectations, vary considerably and can be contradicting. An organization's main goal is to balance stakeholders' needs in a general attempt to satisfy them. A company's primary relation with society includes all the interactions with groups that result in its fulfilling its ultimate goal of producing goods and services for customers. Therefore, stakeholders can be segmented into two main groups. The first group is primary stakeholders, who are considered critical to the organization's survival and actions, and include customers, communities, suppliers, employees, and investors. The second group is secondary stakeholders, groups that are interested in or affected by organizational conduct; these include the general public, media, NGOs, government, and activists groups. However, secondary stakeholders should not be considered in any case less important and in some cases could be primary stakeholders depending on the company sector.

CASE STUDY: CAMPBELL'S 2010 SUSTAINABILITY REPORT EXTRACT

Stakeholder Engagement

Campbell applies multiple strategies to engage our stakeholders and gather diverse external perspectives on our business activities. Campbell uses multiple tools to gather diverse external perspectives on our business activities.

In addition to third-party consumer and customer research, our Consumer and Customer Insights, our Consumer Affairs departments, and our Customer Relationship Managers help us identify the priorities of our marketplace stakeholders.

Specific processes are used to obtain feedback from our suppliers, employees, investors, and external thought leaders in areas including health and nutrition, food safety and quality, environmental stewardship, community relations, and employee engagement. We also conduct benchmarking for leadership performance and interact with many trade and issue management groups worldwide. To complement the rich input we receive through our existing stakeholder engagement models, in Fiscal Year 2009 we also conducted specific surveys on CSR and sustainability with consumers, key customers, suppliers, and internal brand managers. Consumers genuinely appreciate the opportunity to provide feedback on CSR. Below are just three of the more than 100 comments we received after our most recent consumer survey of key CSR attributes and perceptions.

Campbell's business leaders serve as trustees or members of the advisory boards of many organizations engaged in the CSR/sustainability agenda, including, for example, the following:

- American Council for Fitness and Nutrition
- Healthy Weight Commitment Foundation
- Agricultural Sustainability Institute, University of California at Davis
- Food Allergy Research and Resource Program
- Center for Food Safcty at the University of Georgia
- Food Allergy & Anaphylaxis Network
- Boston College Center for Corporate Citizenship
- Society of Consumer Affairs Professionals (SOCAP)
- Net Impact
- Food for All
- Association for Consumer Research
- Foundation for Strategic Sourcing
- European Cluster for Fruit and Vegetable Innovation
- Food Bank of South Jersey
- Students in Free Enterprise
- United Way of Camden County
- United Way, Wilton/Norwalk, CT

In addition, Campbell has established a Scientific Advisory Panel to provide external expertise on vegetable nutrition. The panel has been designed to provide strategic insights on current and emerging science on vegetable and plant ingredients including their nutrient/phytonutrient composition, the health benefits that these vegetable/plant ingredients and phytonutrients could provide, how manufacturing and processing may affect these nutrients and phytonutrients, and how product attributes can be delivered from this science. All of this information is part of the materiality assessment that is conducted annually in the CSR and sustainability strategic planning process [4].

STAKEHOLDER RELATIONS AND ATTRIBUTES: POWER, LEGITIMACY, AND URGENCY

Evidently, firms and stakeholders have a relationship most of the time based on power dependence. Some definitions focus on the firm's dependency on its stakeholders in order to survive, while others focus on the stakeholders' dependency on the firm for upholding their rights, minimizing problems, or achieving their interests. In general, broader definitions emphasize stakeholders' power to influence the firm's conduct, irrespective of legitimate claims. In order to better illustrate the issue, let us define power. Power is "the probability that one actor within a social relationship would be in a position to carry out his own will despite resistance" [5]. Therefore, a stakeholder group has power as it has or can obtain access to coercive, utilitarian, or normative means in order to impose its will in the relationship with the organization. Coercive power is based on the physical resources of force, violence, or restraint; utilitarian power is based on material or financial resources; and normative power on symbolic resources.

Another stakeholder attribute is legitimacy, which refers to "the perceived validity or appropriateness of a stockholder's claim to a stake" [6]. Consequently, owners, employees, and customers represent a high degree of legitimacy due to the nature of their relation with the organization, which can be characterized as explicit, formal, and direct. The last attribute is urgency, which refers to "the degree to which the stakeholder's claim calls for the business's immediate attention or response" [1]. Based on these three attributes, Mitchell et al. [6] created a typology of stakeholders, which depicts the degree towards which managers must attend stakeholders based on their assessment, and of competing stakeholder claims using the criteria of legitimacy, power, and urgency.

CASE STUDY: INTEL 2010 SUSTAINABILITY REPORT EXTRACT

Stakeholder Engagement

We derive significant value from our diverse stakeholders and maintain formal management systems to engage with, listen to, and learn from them. We take their feedback seriously and, when appropriate and relevant to our business, incorporate it into our thinking and planning. We prioritize our stakeholders and their concerns by looking at both the relevance of the stakeholder's relationship to our business and the importance of the issue being raised. We evaluate our community programs based on local input, and we work to adapt our reporting methodology and the content of this report to meet the needs of our stakeholders.

We have developed a number of tools and processes that provide valuable, ongoing feedback on our performance and strategy. In addition to face-to-face meetings, we generate discussion through Web tools and social media. We maintain an e-mail account on our corporate responsibility website that enables stakeholders to share their issues, concerns, and comments directly with members of our corporate responsibility team. Through this account, we receive and respond to hundreds of messages each year on a wide variety of topics. In addition, we have an external CSR@Intel blog, where members of our corporate responsibility team and leaders across Intel discuss their views and opinions and receive and respond to comments made by other blog participants. In 2010, we continued to expand our use of other social media channels, such as Twitter, to reach new audiences with information on our corporate responsibility performance.

For more than 10 years, we have completed an annual "socially responsible investor (SRI) road trip" to meet with leading environmental, social, and governance research firms and SRIs to review our Corporate Responsibility Reports, gain a better understanding of emerging issues, help set priorities, and gather feedback on our performance. In 2010, we met with representatives of more than 20 firms in three cities. Key discussion topics included water conservation, political accountability, conflict minerals and supply chain responsibility, and reporting best practices.

We also held a stakeholder panel to gain input for the development of our 2020 environmental goals. This session, convened by Business for Social Responsibility, included investors, peer companies, and representatives from NGOs. Priority issues identified in this session included water, climate change and energy conservation, transparency and reporting, supply chain responsibility, and using Intel technology to address environmental challenges. Feedback from these meetings has informed improvements in our reporting and goal-planning processes.

We work with community stakeholders to consider the impact of our operations at all phases: entering, operating, and exiting. When entering a community, we work with third parties to conduct needs assessment studies to prioritize our community engagement activities. We also begin working with community organizations to develop programs and initiatives prior to commencing operations. When making the difficult decision to close a facility, we try to minimize the impacts on employees and the local community by collaborating with local officials and providing severance packages and job search support for employees.

During our operating phase, we work to build relationships with local stakeholders through informal meetings, community advisory panels (CAPs), working groups, and community perception surveys (usually completed by third parties). CAP members provide constructive input on a broad range of issues, such as education, environmental impact, health and safety, and emergency response and management. For example, the Intel New Mexico Community Environmental Working Group (CEWG) meets monthly to discuss concerns about Intel's environmental impact. The CEWG is chaired by John Bartlit, chairman of New Mexico Citizens for Clean Air and Water, and is facilitated by a third party. CEWG meeting minutes and agendas for the last six years are posted on the CEWG website. In 2010, we also conducted third-party-administered community focus groups and surveys to understand the concerns and priorities of the local community. In early 2011, we launched a pilot in New Mexico for a new transparency website, "Explore Intel," which provides real-time disclosure, monitoring, and videos for the local community. The website also features a blog and e-mail account where community members can engage with our environmental managers (Intel CSR Report, p. 12).

BALANCING STAKEHOLDERS' EXPECTATIONS

Once the organization successfully identifies and classifies its stakeholders, it must then identify the needs and expectations of the stakeholders. Those expectations and needs are often expressed in various forms, such as letters of complaint or direct dialog. However, sometimes there are needs and expectations that are more difficult to define, as they are not expressed in a direct way or even not ("expressed") communicated at all. However, the major question is not how to identify those needs and expectations, but rather how to address their conflicting nature.

Conflicting needs and expectations exist among all stakeholder groups. For example, the desires of internal stakeholders may differ from those of external stakeholders. For example, customers want low-price, high-quality, and excellent service. On the other hand, employees want high wages, high-quality working conditions, and fringe benefits, while suppliers want low risk and high rate of returns; communities want companies to benefit the community at large, increased local investment, and employment, and the list goes on. Evidently, stakeholder relations are more multifaceted than simple PR work and must be tailored to meet individual stakeholder requirements. These procedures are as diverse as the list of key stakeholders and should take into account each party's appetite for direct involvement, preferred communications mechanisms, and key motivators.

In modern dynamic environments, one should not underestimate the value of strong stakeholder relationships, given the power of external influences to either enhance or disrupt organizational stability and viability. The benefits of good stakeholder relations are usually intangible and therefore hard to quantify. However, they do offer some form of insurance in an uncertain world. Furthermore, opening new channels of communication can offer fresh insight and value, which would have been otherwise inaccessible.

CASE STUDY: EXXONMOBIL'S 2010 CORPORATE CITIZENSHIP REPORT ENGAGEMENT

Like most energy providers, we operate at the center of challenging issues that require transparency and responsiveness. Our increasingly diverse group of stakeholders includes governments, NGOs, communities, shareholders, customers, suppliers, employees, and others. External engagement is fundamental to how we conduct our business.

Ongoing dialog and engagement with our stakeholders provide opportunities to listen to concerns, identify material issues, and benchmark performance against expectations. We focus our engagement efforts primarily on groups and individuals directly impacted by, or who have a direct impact on, our operations. Our engagement takes many forms, including internal and external one-on-one and group dialogs and briefings; senior executive speeches; quarterly earnings teleconferences; focus groups; community consultations; e-mail communications; publications, such as the Corporate Citizenship Report, Summary Annual Report, The Outlook for Energy, and ExxonMobil Perspectives; and content on our website.

It can be challenging to address certain stakeholder issues that directly conflict with our business practices. Additionally, it is not uncommon for us to encounter opposing stakeholder views. For example, many diverse viewpoints exist surrounding shale gas development using hydraulic fracturing. We listen to all perspectives on the energy debate and consider these discussions in long-term planning.

BENEFITS FROM THE CORPORATION'S RESPONSIBLE BEHAVIOR: THE CSR DEBATE

CSR performance can be extremely beneficial for enterprises in many different ways. Although some companies may be forced to join the new movement of social and ethical activities, the long-term multi-effectiveness of sustainability seems apparent.

Some of the primary benefits are:

1. *Improved Financial Performance*   There was a long debate among investment and business communities on whether there is a real connection between socially responsible business practices and positive financial performance. The results of several studies show that improved financial performance seems to be undisputable. Companies have a sustainability strategy if someone considers the fact of 100% transparency of financial records and awareness for each financial change on behalf of all parties. This creates a more precise economical management as well as more trust-based financial exchanges, both factors that lead to improved economic results.

2. *Reduced Operating Costs*   Many of the sustainability practices and initiatives, such as those that address the environment or workplace, may lead to significant cost savings. For example, efforts that aim to reduce climate-changing greenhouse gases also increase energy efficiency and reduce unity bills. In addition, many recycling initiatives cut waste disposal costs and generate income by selling recycled materials. Although these examples cover only 30% of the possible sustainability initiatives, they account for 70% of cost reductions.

3. *Added Value for the Product Leading to Customer Loyalty*   In a global market where competitive products and services are becoming more homogeneous on price and quality, a socially responsible profile can serve as the decisive element of customer choice. The initiative for more socially and environmentally oriented production results in products and services that go beyond the customer's perceived value.

4. *Enhanced Brand Value and Reputation*   At times, sustainability by businesses positively influences the public. Even though a corporate donation or an environmentally driven action might be expense for the business, it seems to establish long-run brand image. Brands are the expression of the corporate value system and as such the company's reputation. A brand no longer represents simply the finished product or service; rather it is the culmination of all aspects of business, including how and with whom it does business and the influence it can have on

external and internal parties. Thus, customers are increasingly drawn to brands that have good reputations in sustainability-related areas. A company considered responsible can benefit from both its enhanced reputation with the customers and its reputation within the business community, increasing a company's ability to attract capital and trading partners.

5. *Increased Worker Commitment Leading to Employee Loyalty* CSR or Sustainability can act as a very powerful motivator for the staff. Employees that are given the opportunity to express their individuality and give something back to the community tend to be happier with their daily work and are more enthusiastic about their company. Moreover, offering employees the freedom to collaborate on particular projects can promote teamwork and relationships between employees and superiors and create a healthier and more productive work environment.

   According to a research conducted by the British corporate think tank, *The Work Foundation*, companies can improve their chances of hiring and retaining talented staff if they make efforts to recognize and support employees' needs and pay more attention to environmental and community concerns:

   (a) The group's study of more than 1000 British workers concluded that about one in ten is an ethical enthusiast or strong supporter of CSR practices. These groups of workers tend to fall within the age ranges of 18–24 or 45 and over.

   (b) The study also found that employers with no concern for corporate ethics are more likely to lose staff within a 12-month period.

   (c) The deputy director of research at The Work Foundation, Stephen Bevan, highlights the misconception that most employers assume that the employees resign due to financial reasons. However, according to Bevan, only about 10% of employees leave because they are not sufficiently satisfied with their pay package.

6. *Good Relations with Government and Communities* Another way sustainability benefits the company is through the establishment of mutual trust and respect between business and local communities. When the company seeks to improve society and acts not only for its own profit, it gains more privileges. As a result, the bond between communities and the private sector is strengthened.

Possible advantages for the enterprise:
   (a) Gains the community support
   (b) Creates alliances with community sectors

(c) Gains the political decision-makers' confidence

(d) Fosters relationships with other businesses that might become clients

THE CSR DEBATE

After analyzing the various advantages of adopting CSR strategies, it is still necessary to mention that there are arguments concerning the responsibilities of a business, as a balance of power with responsibility. It is essential to manage equilibrium between power and responsibility. Today companies have unprecedented power, which may harm society if those who are in charge do not follow specific steps. However, if firms use their power in a responsible and trustworthy manner, the risks of poor judgment and shortsighted operations shall be eliminated. Businesses committed to social responsibility are aware that if they abuse the power they have, they might lose it permanently.

REFERENCES

1. Carroll AB, Buchholtz AK. *Business & Society: Ethics and Stakeholder Management*. Cincinnati, OH: South-Western Pub. Co.; 2003.
2. Freeman RE. *Strategic Management: A Stakeholder Approach*. Boston, MA: Pitman; 1984.
3. Clarkson M. *A Risk Based Model of Stakeholder Theory*. Toronto: Centre for Corporate Social Performance & Ethics, University of Toronto; 1994.
4. Campbell CSR Report; 2010, p. 27–28.
5. Weber M. *The Theory of Social and Economic Organization*. New York: Free Press; 1947.
6. Mitchell RK, Agle BR, Wood DJ. Toward a theory of stakeholder identification and salience: defining the principle of who and what really counts. Acad Manage Rev 1997;22(4):853–886.

CHAPTER 16

Sustainability Reporting

SUSTAINABILITY REPORTING

There are two main reasons that make sustainability reporting vital. Firstly, all over the world there are sustainable development challenges, and in order to bring solutions, companies must play a significant role by promoting economic, environmental, and social reporting. On the other hand, corporate scandals resulted in the loss of trust in the businesses, which can be overcome through a more open and clear communication.

CONTEXT OF REPORTS

The reporting process involves seven topics that corporations are interested in. The first one is called Financial Analysts. The main point of this topic is to find ways to connect the company with the financial world. Many corporations use the sustainability report in order to gain ground and build a stronger profile among financial analysts. Into this sector of the sustainable reporting, companies provide the shareholders with information about the money the company spends for expanding, operating, and protecting the environment.

The second topic that sustainability reports incorporate is called Assurance and Verification. This topic includes detailed reporting process by specific companies—called verifiers—that inform the shareholders for the future goals of the company and recommendations in order to accomplish them.

The topic that follows is called Supply Chain Reporting and discusses the complexity of cooperating with the suppliers. Nowadays, there is an increasing percent of stakeholders who take into consideration the supply chain. For this

Practical Sustainability Strategies: How to Gain a Competitive Advantage,
First Edition. Nikos Avlonas and George P. Nassos.
© 2014 John Wiley & Sons, Inc. Published 2014 by John Wiley & Sons, Inc.

reason, companies must be very critical when deciding about the suppliers. Ford Motor Company faced a serious problem when more than 100 people had died in accidents due to problems in the tires of the vehicles—supplied by Firestone. Both customers of Ford Motor Company and Firestone lost their trust in these companies.

The fourth topic that is discussed in the sustainability reports is called Emerging and Transition Economies and refers to the way in which company policies contribute know-how and experience in emerging economies.

The Economic Bottom Line is the fifth issue that the sustainability reports are communicating about. This topic goes beyond financial accounting and discusses positive or negative economic social and environmental impacts that companies have on their stakeholders. The whole concept is to go beyond balance sheets into Economic Value Added metrics in order to define clearly their true meaning. For example, Rio Tinto in its sustainability report informs the company's stakeholders of its economic contribution to the local communities.

The topic that follows is called Brands and Reputation. This subject discusses how corporate reporting is related to the reputation, the brand name, and the values of the company. Reputation, brand name, and values of the company are considered to be extremely important in order for a company to be successful. The thing that makes brand name essential is the trust that it carries. For example, Nike, McDonald's, and Coca-Cola are companies that people trust because of their commitment to their values.

The last topic that sustainability reports take into account is about Governance. Corporate governance discusses how the company's management deals with economic and social issues. Corporate governance is concerned with creating a balance between economic and social goals and between individual and communal goals while encouraging efficient use of resources, accountability according to the interests of individual's corporations and society.

CHANGES OVER THE YEARS

During the earlier years of the 2000s, only a few dozen companies issued sustainability reports. Furthermore, sustainability reports focused rather exclusively on environmental protection and stewardship. Yet, as the years progressed, sustainability reports grew more comprehensive and adopted a holistic approach, investigating social and economic concerns as well following global guidelines such as Global Reporting Initiative (GRI).

Reports issued after the new millennium provide shareholders with much more information and insight into a company's operations. This shift gained traction as companies began to recognize the importance and utility of fostering a socially, economically, and environmentally responsible image for shareholders and stakeholders alike.

The two following case studies achieved the highest GRI G3 version score available, A+. The score A denotes that the sustainability report achieved the highest level of completeness and comprehensiveness, while the "+" symbol signifies a third party assured the report's content and reporting processes. This descriptive indicator communicates to stakeholders the thoroughness of the company's sustainability report and thus indirectly reflects how progressive or mature a company's sustainability operations are. The following are extracts from Hess and Dow's 2010 [1, 2] sustainability reports. These extracts discuss the reporting parameters and verification processes adopted by the companies, which enabled them to achieve the A+score. The third section of GRI-structured sustainability reports typically addresses these issues.

HESS CORPORATION 2010 SUSTAINABILITY REPORT

Reporting Standards

At Hess Corporation we report our sustainability performance based on the GRI G3 guidelines, to an A+reporting level. Our report is also based on the International Petroleum Industry Environmental Conservation Association (IPIECA), American Petroleum Institute (API) Oil and Gas Industry Guidance on Voluntary Sustainability Reporting, the 10 Principles of the United Nations Global Compact (Global Compact), and industry best practices.

The GRI Content Index included near the end of this report summarizes the completeness of our reporting. Detail is provided according to GRI G3 indicators, which are cross-referenced with IPIECA indicators and the Global Compact.

Boundary Setting

The principal facilities and assets operated by Hess Corporation and its subsidiaries and joint ventures during calendar year 2010 are included in this report. Data presented in this report refer to gross figures from operated facilities, joint ventures where we have significant influence, according to the GRI framework, and third-party activities where Hess has overall responsibility as specified in contractual arrangements. For Samara-Nafta Operations, we include net equity, greenhouse gas (GHG) data, and social investments spending.

To facilitate comparisons with prior annual corporate sustainability reports, joint venture data for SonaHess (Algeria) and the Carigali Hess Malaysia/Thailand Joint Development Area Block A-18 (MTJDA) are included in Hess operated totals.

Some quantitative environment, health, and safety data are reported on a normalized basis to facilitate year-on-year comparisons. HOVENSA social investments, health and safety, and certain environmental data are provided separately.

We report GHG emissions on an operated basis for Hess operated assets, SonaHess and Carigali Hess. Net equity emissions intensity data are provided on a net equity share basis for operated facilities, joint ventures including HOVENSA, and non-operated facilities in which we hold an interest.

Internal Quality Assurance

We have documentation and information systems in place to ensure consistent and reliable data collection and aggregations from all of our Hess operated and joint venture assets. We conduct corporate- and business-level Quality Assurance/Quality Control (QA/QC) reviews and validation to evaluate the accuracy and reliability of facility specific and aggregated data.

Restatements and Additions

The 2009 fines and penalties in the performance data table have been restated to reflect payment of the Port Reading OSHA National Emphasis Program (NEP) fine of $97,500, paid on October 20, 2009. The units for several of our metrics have been changed this year to reflect IPIECA's Oil and Gas Industry Guidance on Voluntary Sustainability Reporting (2010) and industry sector best practices.

Report Availability

Print copies of our sustainability report are distributed to our employees and external stakeholders and are available upon request. The report is also posted on the Hess (www.hess.com), GRI (https://www.globalreporting. org/Pages/default.aspx), and Corporate Register (www.corporateregister. com) websites and is uploaded to the United Nations Global Compact website (www.unglobalcompact.org) as our annual Communication on Progress.

DOW 2010 GLOBAL REPORTING INITIATIVE REPORT: THE ANNUAL SUSTAINABILITY REPORT

Report Parameters

Reporting Period   Based on 2010 corporate data for the year ending December 31, 2010.

Date of Most Recent Previous Report   This is the eighth GRI Sustainability Report. The previous report covered 2009 and can be found at our Sustainability Reporting site.

REPORT SCOPE AND BOUNDARY

Process for Determining Report Content

Two complementary processes merged to form our picture of materiality, as defined in GRI Reporting Guidelines. The first is the development of and the attention given to the implementation of our 10-year Sustainability Goals. The second is an annual Public Policy Issues Prioritization process. The 2015 Sustainability Goals cover the majority of our material issues and are illustrated in shaded boxes. The annual Public Policy Issues Prioritization process identifies additional items of importance to the company and to its stakeholders.

The seven goals represented by orange shading were declared in 2006, with specific targets to drive improvements by 2015. These remain a focus for guiding our pursuit of being a sustainable company. Links on the names of the goals will take you to more information regarding each goal located within this report.

Energy and Climate Policy are high in significance of impact and importance to stakeholders because Dow operates at the nexus between energy and all the manufacturing that occurs in the world today. Over 96% of the products made have some level of chemistry in them, so no one has more at stake in the solution—or more of an ability to have an impact—on the overlapping issues of energy supply and climate change than we do.

Sustainable chemistry is a commitment to increase the percentage of sales that are based on highly advantaged chemistry to 10% by 2015, up from 2% in 2007. These more sustainable products are produced using renewable feedstock, more environmentally friendly raw materials, less energy or reduced footprint, or some combination thereof. Increasingly, our customers and our customers' customers are interested in purchasing more sustainable products.

Also displayed in the high importance and high impact category are the Local Protection of Human Health and the Environment (LPHHE) and Community Success Goals. Both of these 10-year Sustainability Goals highlight our commitment to making Dow a safer place to work and to improving our relationships in the regions where we operate, ensuring that we are a contributing and valuable partner in making the area a quality place to work, live, and raise a family. The LPHHE Goal tracks our commitment to best-in-class performance in keeping people safe and healthy as well as reducing our footprint. The Community Success Goal defines what stakeholders view as of greatest importance for the company to contribute to the quality of life in the region.

The Breakthrough to World Challenges Goal is unique. It challenges us to seek out and address world-scale challenges. We pledge to use science and technology—especially in areas where we are leaders—to make the world better for future generations. One example of a significant breakthrough is Omega-9 oils. These healthier oils for cooking and food processing are contributing to healthier lives around the world (see our Q2 2010 Goal Update for more information about how the replacement of trans fats from food may help eliminate one in five heart attacks in the United States).

The company has recognized risks related to water quality and availability for more than 10 years. In 2008, at the World Economic Forum in Davos, Switzerland, the company's CEO led a session to foster broader recognition and stimulate a growing role for business to address water's connection to human health, to support agricultural production, and for commercial applications.

The driver of significance for many stakeholders is the growing awareness that over a billion people do not have access to clean water. The Dow Water and Process Solutions business is advancing the efficiency of multiple technologies to address water health issues and help prevent the advance of water stress in geographies where supply is currently threatened or will be as demands continue to grow (http://www.dowwaterandprocess.com).

Biodiversity and Ecosystem Services are gaining recognition as extremely valuable services naturally occurring in nature, which have been historically available for little cost. Dow, in collaboration with The Nature Conservancy, is working to develop, implement, and refine ecosystem services and biodiversity assessment modeling tools on Dow sites and in Dow business decisions. Chemical Security is of high importance to stakeholders. Our modern society and global economy depend on substances that if not handled, stored, transported, or secured properly, could cause harm. These substances serve as critical components for medicines, plastics, automobiles, and building materials to name just a few. In today's world, there is concern that chemical facilities could be attacked by or those who wish to intentionally cause harm, disrupting our way of life, could use chemicals. Read more about how we take an active role in this important area at http://www.dow.com/sustainability/safety.htm.

The goal of delivering several commitments in the area of Product Safety is illustrated in the middle of importance and impact. The on-track implementation to publish a Product Safety Assessment (PSA) for every Dow product by 2015 is setting a new standard within the chemical industry. We further commit to addressing hazard and exposure gaps, aligned with the ICCA Global Product Strategy Risk Assessment System, on all of our products or product families by 2015, in line with our sustainability goals. We make PSAs accessible to the public at http://www.dow.com/productsafety/.

Reports and Investing

As described, sustainability reporting has exploded over the past decade. While there are a variety of reasons for the growing popularity of sustainability reporting, such as the improvement or maintenance of a desirable public image, these documents also serve to attract and persuade investors. First, investors concern themselves with sustainability reports because the publication of an annual report reflects a strong and organized managerial structure. The United Kingdom's news outlet, *The Guardian*, explains, "The non-publication of a report, or the absence of published policies, targets and performance data, is increasingly likely to be taken as evidence that the company does not recognize social and environmental issues as management priorities, thereby raising wider questions about the quality of the company's risk systems and processes" [3]. This is especially true for Fortune 500 companies. In short, a business that exhibits sound and effective management strategies appears much more attractive to potential investors.

Changes in the values of individual investors have also contributed to the recent explosion of sustainability reporting. As investors grow more sensitive to social and environmental issues, sustainability performance indicators will have a greater impact on investment decisions. Naturally, this is a generalized statement as not all investors concern themselves with social and environmental issues. Nevertheless, the growing public sensitivity to these topics appears to be influencing investment decisions. An increasing number of investors are making commitments to integrate consideration of social and environmental issues into their investment decision-making processes, and better quality environmental and social performance data are increasingly available. It is therefore to be hoped that they will, over time, see investors paying greater attention to companies' corporate responsibility performance [4].

Sustainability reports not only record important advancements in corporate environmental and social strategies but also foster a competitive environment where companies are compelled to match the significant benchmarks of others. This concept is important for investors as increased competition may accelerate financial performance. Companies that choose to enter this competitive realm stand to gain, while those that neglect sustainability reporting may be left behind.

Report Utility and the Investing Debate

However, it is important to note that there is a debate concerning the degree to which sustainability reporting matters to investors. Those skeptical of the utility of sustainability reporting argue that most investors are strictly concerned with financial growth over a very short period of time, usually less than two years. These critics contend that while the publication of sustainability reports may serve to cultivate public trust and engagement in the long term, investors are solely concerned with short-term financial growth; they argue the latter is not related to sustainability goals and may even come at the expense of social and environmental responsibility.

Dr. Rory Sullivan [5] of *The Guardian* explores this perspective. He writes, "It is common for investors to focus their attention on what they see as the most important factors that will drive business performance over a relatively short time period (typically somewhere between a few months and two years). Over this sort of time frame, the vast majority of environmental and social issues are, unfortunately, not financially significant" [4]. According to this opinion, although investors may care about environmental and social issues on a personal level, these concerns are not factored into their investment decisions.

Sustainability Indexes

Despite debate, it is clear that many investors do consider sustainability performance when they choose to invest. This position is supported by significant changes in the ways global investment organizations, such as Bloomberg, Dow Jones Sustainability Index (DJSI), and FTSE4Good, value and incorporate sustainability metrics. For example, the development of the DJSI is one of the most notable manifestations of such changes. In order to allow investors to make more informed market decisions on sustainability performance, Dow Jones launched the DJSI in 1999. Today, there are many DJSI licenses held by asset managers; these licenses manage more than eight billion USD [6] worldwide. According to Dow Jones, the utility of this index is twofold: "The DJSI enable investors to integrate sustainability considerations into their portfolios while providing an effective engagement platform for encouraging companies to adopt sustainable best practices" [7].

This DJSI offers further evidence to support the positive relationship between sustainability and financial performance. Dow Jones explains, "A quantitative analysis of the historical sustainability information compiled by SAM over the years, and upon which the DJSI are based, reveals that an investment strategy that selects sustainability leaders and avoids sustainability laggards contributes to superior long-term investment results with improved risk/return profiles" [7]. Such evidence serves to further accelerate the incorporation of sustainability performance into investment decisions and management systems and bodes well for the future of sustainable progress.

REFERENCES

1. Hess Corporation 2010 corporate sustainability report. Hess Corporation, New York. Available at http://www.hess.com/reports/sustainability/US/2010/default.pdf/. Accessed 2013 Jul 3.

2. Dow J. 2010 global reporting initiative report: the annual sustainability report. Available at http://www.dow.com/sustainability/pdf/233-00864-GRI-2010.pdf. Accessed 2013 Jul 3.

3. Sullivan R. Should companies produce corporate responsibility reports? *The Guardian*, March 2011. Available at http://www.guardian.co.uk/sustainable-business/blog/corporate-responsibility-reports. Accessed 2013 Jul 3.

4. Sullivan R. What do investors want to know about corporate responsibility, *The Guardian*, March 2011. Available at http://www.guardian.co.uk/sustainable-business/investors-corporate-responsibility?INTCMP=SRCH. Accessed 2013 Jul 3.

5. Sullivan R. The changing face of corporate sustainability, *The Guardian*, March 2011. Available at http://www.guardian.co.uk/sustainable-business/investors-corporate-responsibility?INTCMP=SRCH. Accessed 2013 Jul 3.

6. Abu AA. Why we need sustainability reporting, *World Finance*, October 2011. Available at http://www.worldfinance.com/strategy/investor-relations/why-we-need-sustainability-reporting/. Accessed 2013 Jul 3.

7. Dow J. DJSI brochure. Available at http://www.sustainability-indices.com/library/publications.jsp. Accessed 2013 Jul 3.

CHAPTER 17

Metrics for Sustainability

Sustainability metrics are a combination of economic, social, and environmental indicators related to performance and stakeholder perceptions. The term "metric" encompasses any way of measuring performances, which can be adopted during different stages of the decision-making process, in order to minimize the impact of the decision, as well as after, so as to reduce and improve the actual situation.

It is hard to develop a metric, which can be relevant for all sectors or situations, and therefore a unique metric for sustainability. In the last years particularly, there has been a flourishing of different certifications and standards, concerning all possible issues within sustainability. As proof of this there are numerous and different metrics that can be used to quantify sustainability and, most of the time, just some of its aspects. Some of the most recognized worldwide are the GRI standard and the ones released by the International Organization for Standardization (ISO) (ISO 14000, 14046, 14006).

Measuring sustainability is one of the (endless) challenges that companies, as well as other types of organizations, are facing nowadays. Since the concept of sustainability is not yet transferable into quantitative data, the consequence of this is a higher uncertainty concerning the method to apply, and this, until now, has facilitated the proliferation of tools to measure different aspects of sustainability, often with different methodologies.

Since the road to sustainability is a never-ending process, it is important to measure sustainability because this will allow organizations to keep track of the improvement achieved up to a certain point. Indeed, different metrics can be used to evaluate the progress made and set new parameters and targets to be achieved; this will also give more credibility of action in front of stakeholders. "Monitoring and metrics have always been essential components of closed physical systems. They are integral to the scientific method.

Practical Sustainability Strategies: How to Gain a Competitive Advantage,
First Edition. Nikos Avlonas and George P. Nassos.
© 2014 John Wiley & Sons, Inc. Published 2014 by John Wiley & Sons, Inc.

In this context each indicator should have a threshold and a target to guide political and social action" [1].

METRICS IN THE GRI GUIDELINES

One of the most adopted global guidelines for sustainability is the GRI's Sustainability Reporting Guidelines. The G3.1 version contains metrics, in regard to human rights, local community impact, and gender. The document is divided into two parts: the first in which the guidance on how to report is presented and the second that clarifies what should be reported as disclosures on management approach and as performance indicators.

PART 2: STANDARD DISCLOSURES

- Strategy and profile
- Management approach
- Performance indicators

Furthermore, the performance indicators are organized into three categories: social, environmental, and economic, where the social is divided further into labor, human rights, society, and product responsibility [2].

According to GRI guidelines the Materiality Principle is a method of determining which indicators should be reported in a GRI Sustainability Report and should be the focus of the organization for managing, measuring, and reporting.

The GRI Guidelines Materiality Test shown in Figure 17.1 should evaluate the importance of monitoring the indicators with regard to the organization's current and future performance.

CASE STUDY ABM SUSTAINABILITY REPORT (2011)

ABM has produced our Fiscal Year 2011 Sustainability Report for clients, shareholders, employees, supplier partners, and others with an interest in our corporate approach to sustainability. With 103 years of facility services experience, we continue to serve thousands of clients across the United States and in various international locations. Over the next couple of pages, Henrik Slipsager, ABM president and chief executive officer, will detail today's ABM, including our recently launched new brand and the company's transformation towards our new vision: to become the global leader in Integrated Facility Solutions.

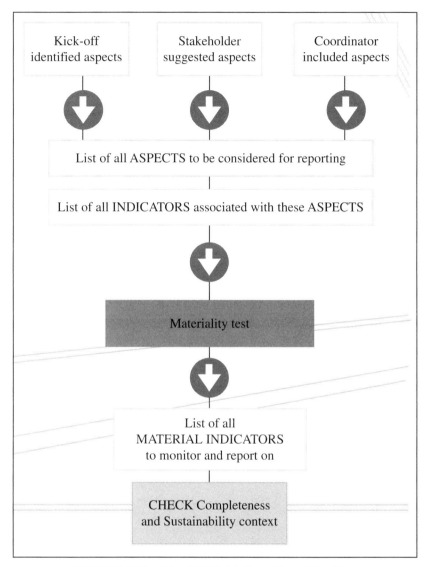

FIGURE 17.1   The GRI Guidelines Materiality Test.

We have proclaimed that it's a "New Day" at ABM, a momentous time that most certainly includes an unwavering commitment to environmental stewardship. We understand that as a corporate citizen and industry leader, we must minimize our environmental impact across each of our business units while providing world-class services to our clients. We are keenly aware of our impact upon the fiscal, social, and physical environments we all share, and our operations, both internally and externally, reflect that mindset. This sustainability report details the sustainability achievements, goals, and challenges that ABM has identified within our operations, and it publicly identifies our commitment to monitor and improve our impact on people and the Earth. ABM is a publicly traded company and thus held to a higher standard of ethics and compliance compared to many competitors in the facility services industry.

A critical example of ABM's compliance is our Sarbanes Oxley (SOX) certification. ABM rigorously evaluates its control environment throughout the year for each of its business units, ensuring that our financial operations and reporting processes are well designed, effective, and efficient. Sustainability reporting is an ever-changing discipline, and as our corporate sustainability programs continue to mature, so too must our reporting. We strive to be at the forefront of best practices, both inside and outside our industry segment.

To that end, this year we have submitted this report for the assessment of the Centre for Sustainability and Excellence (CSE), an independent, outside entity. CSE's evaluation is included later in this report (Table 17.1).

TABLE 17.1   GRI content table

	Standard Disclosures	Section	Coverage
1.	Strategy and analysis		
1.1	Statement fully reported from the most senior decision-maker of the organization about the relevance of sustainability to the organization and its strategy	CEO statement	Fully reported
1.2	Description of key impacts, risks, and opportunities	CEO statement Risks and opportunities Financial implications and risks/opportunities for ABM due to climate change regulation	Fully reported
2.	Organizational profile		
2.1	Name of the organization	Introduction	Fully reported
2.2	Primary brands, products, and/or services	Building value through sustainability ABM operational structure and service offerings	Fully reported
2.3	Operational structure of the organization, including main divisions, operating companies, subsidiaries, and joint ventures	Building value through sustainability ABM operational structure and service offerings Significant changes this year	Fully reported
2.4	Location of organization's headquarters	About ABM	Fully reported

TABLE 17.1 (*cont'd*)

	Standard Disclosures	Section	Coverage
2.5	Number of countries where the organization operates and names of countries with either major operations or that are specifically relevant to the sustainability issues covered in the report	About ABM	Fully reported
2.6	Nature of ownership and legal form	About ABM	Fully reported
2.7	Markets served (including geographic breakdown, sectors served, and types of customers/ beneficiaries)	About ABM	Fully reported
2.8	Scale of the reporting organization	About ABM Data at a glance	Fully reported
2.9	Significant changes during the reporting period regarding size, structure, or ownership	Significant changes this year	Fully reported
2.10	Awards received in the reporting period	Recognition for our integrity and expertise	Fully reported
3.	Report parameters		
3.1	Reporting period (e.g., fiscal/ calendar year) for information provided	Introduction Report parameters	Fully reported
3.2	Date of most recent previous report (if any)	Report parameters	Fully reported
3.3	Reporting cycle (annual, biennial, etc.)	Report parameters	Fully reported
3.4	Contact point for questions regarding the report or its contents	Introduction	Fully reported
3.5	Process for defining report content	Introduction Report parameters	Fully reported
3.6	Boundary of the report	Report parameters	Fully reported
3.7	State any specific limitations on the scope or boundary of the report	Report parameters	Fully reported
3.8	Basis for reporting on joint ventures, subsidiaries, leased facilities, outsourced operations, and other entities that can significantly affect comparability fully reported from period to period and/or between organizations	Report parameters Significant changes this year	Fully reported

(*Continued*)

TABLE 17.1   *(cont'd)*

	Standard Disclosures	Section	Coverage
3.10	Explanation of the effect of any restatements of information provided in earlier reports and the reasons for such restatement	Report parameters Significant changes this year	Fully reported
3.11	Significant changes fully reported from previous reporting periods in the scope, boundary, or measurement methods applied in the report	Report parameters Significant changes this year Sustainability in ABM operations	Fully reported
3.12	Table identifying the location of the standard disclosures in the report	GRI content table	Fully reported
3.13	Policy and current practice with regard to seeking external assurance for the report	Report parameters	Fully reported
4.	Governance, commitments, and engagement		
4.1	Governance structure of the organization, including committees under the highest governance body responsible for specific tasks, such as setting strategy or organizational oversight	Governance structure	Fully reported
4.2	Indicate whether the chair of the highest governance body is also an executive officer	Governance structure	Fully reported
4.3	For organizations that have a unitary board structure, state the number of members of the highest governance body that are independent and/or nonexecutive members	Governance structure	Fully reported
4.4	Mechanisms for shareholders and employees to provide recommendations or direction to the highest governance body	How our approach to sustainability has evolved and how it affects stakeholders	Fully reported
4.14	List of stakeholder groups engaged by the organization	How we approach sustainability How our approach to sustainability has evolved and how it affects stakeholders	Fully reported

TABLE 17.1 *(cont'd)*

	Standard Disclosures	Section	Coverage
4.15	Basis for identification and selection of stakeholders with whom to engage	How we approach sustainability How our approach to sustainability has evolved and how it affects stakeholders	Fully reported

	Economic Performance Indicators	Section	Coverage
EC1	Direct, economic value generates and distributed, including revenues, operating costs, employee compensation, donations and other community investments, retained earnings, and payments to capital	About ABM, annual report (http://investor.abm.com)	Fully reported
EC2	Financial implications and other risks and opportunities for the organization's activities due to climate change	Financial implications and risks/opportunities for ABM due to climate change regulation	Fully reported
EC4	Significant financial assistance received fully reported from government	Data at a glance	Fully reported
EC7	Procedures for local hiring and proportion of senior management hired fully reported from the local community at locations of significant operation	How we value our people	Partially reported
EC8	Development and impact of fully reported structure investments and services provided primarily for public benefit through commercial, in-kind, or pro bono engagement	Supporting our communities	Fully reported

	Environmental Performance Indicators	Section	Coverage
EN6	Initiatives to provide energy-efficient or renewable energy-based products and services and reductions in energy requirements as a result of these initiatives	Sustainability in ABM services	Fully reported

(Continued)

TABLE 17.1 (*cont'd*)

Environmental Performance Indicators		Section	Coverage
EN7	Initiatives to reduce indirect energy consumption and reductions achieved	Sustainability in ABM operations ABM IT continues to improve our efficiency	Fully reported
EN18	Initiatives to reduce greenhouse gas (GHG) emissions and reductions achieved	Data at a glance Sustainability in ABM operations ABM IT continues to improve our efficiency	Fully reported

Labor Practices Indicators		Section	Coverage
LA1	Total workforce by employment type, employment contract, and region	How we value our people	Partially reported
LA2	Total number and rate of employee turnover by age group, gender, and region	Employee training	Partially reported
LA4	Percentage of employees covered by collective bargaining agreements	Employee training	Fully reported
LA7	Rates of injury, occupational diseases, lost days, and absenteeism and total number of work-related fatalities by region	Training and safety awareness	Partially reported
LA13	Composition of governance bodies and breakdown of employees per category according to gender, age group, minority group membership, and other indicators of diversity	Diversity and human rights sustainability	Fully reported

Human Rights Indicators		Section	Coverage
HR3	Total hours of employee training on policies and procedures concerning aspects of human rights that are relevant to operations, including the percentage of employees trained	Employee training Our code of business conduct	Fully reported

TABLE 17.1 (*cont'd*)

	Human Rights Indicators	Section	Coverage
HR6	Operations identified as having significant risk for incidents of child labor and measures taken to contribute to the elimination of child labor	Risks and opportunities	Fully reported
HR7	Operations identified as having significant risk for incidents of forced or compulsory labor and measures taken to contribute to the elimination of forced or compulsory labor	Risks and opportunities	Fully reported
HR8	Percentage of security personnel trained in the organization's policies or procedures concerning aspects of human rights that are relevant to operations	Employee training ABM security services	Fully reported

	Social Performance Indicators	Section	Coverage
SO2	What is the percentage and total number of business units analyzed for risks related to corruption?	Risks and opportunities	Fully reported
SO3	What is the percentage of employees trained on anticorruption policies and procedures?	Employee training	Fully reported
SO5	Public policy positions and participation in public policy development and lobbying	Governance structure	Fully reported

	Product Responsibility Indicators	Section	Coverage
PR6	Programs for adherence to laws, standards, and voluntary codes related to marketing communications, including advertising, promotion, and sponsorship	Responsible marketing practices	Fully reported
PR7	Total number of incidents of noncompliance with regulations and voluntary codes concerning marketing communications, including advertising, promotion, and sponsorship, by type of outcomes	Responsible marketing practices	Fully reported

THE BS 8900

In terms of comprehensive standards and guidance for sustainability, in 2006 the British Standards Institution (BSI) released the BS 8900 Guidance for managing sustainable development. The purpose of this guidance is to provide a helpful tool for organizations in order to set a sustainable plan, and improve it constantly, to shape it to the new challenges and situations that the company will face. The guidance:

- Provides a framework so organizations can take a structured approach to sustainable development by considering the social, environmental, and economic impacts of their organization's activities
- Is applicable to all organizations, in terms of size, type, etc., including civil societies and trade unions
- Makes it easier for organizations to adjust to changing social expectations
- Helps organizations to connect existing technical, social, and environmental standards, both formal (e.g., ISO 14000 series of standards) and private (e.g., the GRI and the AA1000 standards)
- Offers a maturity pathway for the development of the management of sustainable development issues and impacts
- Provides organizations' stakeholders with a useful tool to assess and engage in improving organizational performance
- Contributes to the dialog on the international standard on social responsibility, currently under development [3]

ISO 26000

In regard to standards, one of the most influential authorities on the topic is the ISO [4], which, in November 2010, released a quite expected ISO guideline on corporate social responsibility, the ISO 26000. The aim of such a standard is to provide guidelines to small and big firms, no matter the sector, on how to face and embed social responsibility within the organization [5].

To do so, the document is suggesting addressing seven core subjects and metrics, which are divided into different issues, in order to cover at least the basic issues on social responsibility. Those subjects are:

- Human rights
- Labor practice
- Environment

- Fair operating practices
- Consumer issues
- Community involvement and development
- Organizational governance [6]

ECOLOGICAL FOOTPRINT

The ecological footprint is the virtual surface required to produce and dispose the goods demanded by humans. According to the Global Footprint Network, already from the middle of the 1980s, humanity overshot the world capacity and is virtually consuming more than the world is able to restore [7]. The ecological footprint is a metric that represents how much land surface is necessary to use in order to produce and assimilate the goods that people (usually grouped by nation) are consuming. This value, which currently is 1.5 times the Earth's capacity, is helpful to understand where the main environmental pressure lies in order to reduce our ecological pressure and, therefore, increase our sustainability [8].

Arguably its usefulness, that is, the benefit coming from the calculation of an ecological (as well as the others) footprint, is limited. For this reason the Environmental Department of the European Commission, together with the Institute for Environment and Sustainability, is developing a new footprint tool, called environmental footprint [9].

The aim of the latter is to produce a harmonized methodology to calculate the environmental footprint of products and organizations, also taking into consideration carbon emissions. It is going to be built on the International Reference Life Cycle Data System (ILCD) Handbook, as well as other existing standards and guidance documents, in order to create as comprehensive and complete a tool as possible (European Commission Environment, 2011) [10].

Another relevant guidance, released by the BSI, is the quite spread out ISO 14000 on environmental management, as well as two interesting ISO, the 14046 and 14067, respectively, on water and carbon footprint.

The ISO 14000 provides guidelines for the implementation of an environmental management system, particularly the 14001 and 14004. The aim of the first is to provide the requirements, while the second provides general guidelines. Furthermore, the other standards focusing on specific environmental issues, including labeling, performance evaluation, Life Cycle Analysis (LCA), communication, auditing, and how we already mention on carbon and water footprint [11].

These last two probably deserve particular attention, about the issues, which they are dealing with. The ISO 14046, or Water footprint—requirements and guidelines, aims to complement the existing standard on LCA and on carbon footprint, as well as of providing clear guidelines on an issue where there are plenty of methodologies. The purpose of this standard is to contain

- Principles and guidelines for a water metric of products, processes, and organizations
- Ways as to how the different types of water (green, blue, and grey) should be considered, as well as the socioeconomic and environmental issues
- Ways of communicating water footprint
- Compatibility with the other ISO 14000 standards [12]

The ISO 14067 is strongly based on the ISO 14044, which was published in December 2009, but is also a significant review of it. The aim of the ISO 14067 is to use the LCA to calculate the carbon footprint of products. It is divided into two parts: the elaboration of a carbon footprint study, ISO 14067-1, and the elaboration of a dedicated carbon footprint communication, ISO 14067-2.

The methodology of the ISO 14067-1 includes all sources of GHG, which goes from carbon storages, emissions from the electricity use in the life cycle (cradle to gate), aircraft transport, and also land-use change. The standard has been designed just for accounting the GHG; however, since it is based on the ISO 14044, it may also include any kind of impact [13, 14].

METRICS FOR CARBON FOOTPRINT

The GHG Protocol was developed by the World Research Institute, together with the World Business Council for Sustainable Development, in October 2010. This document is one of the most adopted methodologies to calculate the carbon footprint of organizations, including the supply chain. Indeed its main aim is to identify risks and possibility of reduction along the supply chain and define improvements. The methodology used for assessing is the LCA, which is adopted for all the stages (therefore is theoretically applicable for other environmental aspects).

The protocol is divided mainly into two documents, the Corporate Accounting and Reporting and the Project Accounting Protocol and Guidelines. The purpose of the Corporate Accounting and Reporting is to:

- Standardize the approach to a representative account of the GHG emission of the organization
- Simplify and reduce the costs of the inventory
- Provide business with suggestion on how to successfully manage and reduce GHG emissions

- Increase the consistency and the transparency in GHG accounting and reporting

The key aspects of the Project Accounting Protocol and Guidelines provide specific indications, ideas, and methods on how to quantify, report, and reduce the GHG. Furthermore, some addressed protocols per sector, like the one for agriculture and public sector, have been released [15].

LIFE CYCLE ANALYSIS

The LCA is a holistic methodology, which can be applied to any kind of impact (environmental, economic, and social), since it takes into consideration all the steps of the production cycle as well as the consumption and the end of life. Generally the steps considered are:

- Extraction and treatment of raw materials
- Educational tools
- Product manufacturing
- Transport and distribution
- Product use
- End of life

The aim of this methodology is to help identify the main pressure and issues within the life cycle and help the decision-makers to mitigate them. The methodology of the LCA is defined in the ISO 14040 standard and consists mainly of four phases:

- Setting of goal and scope (define the methodological framework)
- List of all the inputs and outputs of the product system
- Assessment of the potential impacts related to these inputs and outputs
- Interpretation of the inventory data and impact assessment results related to the goal and scope of the study [16]

CASE STUDY: HEINEKEN GROUP CARBON FOOTPRINT

The Carbon Footprint Model that we have designed calculates the total environmental impact expressed in carbon dioxide output. The model covers, country by country, the whole supply beverage chain, both upstream and downstream, covering all carbon dioxide outputs from crop cultivation

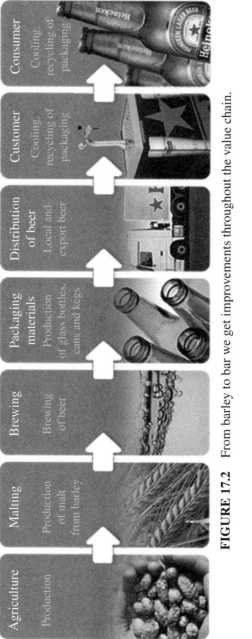

FIGURE 17.2 From barley to bar we get improvements throughout the value chain.

through to the disposal of packaging materials, distribution, and cooling in fridges or draft beer installations. To create an accurate gauge, data that was previously not measured had to be collected by the countries involved and in some cases estimates had to be sourced and created. This has allowed us to map the carbon footprints of our markets in Western Europe and Central Eastern Europe. We are now expanding the scope of our focus to Africa and Latin America. In 2011 we defined how we should incorporate recycling in our carbon footprint calculation (Fig. 17.2).

BALANCED SCORECARD

To manage and integrate all these different standards within the organization, it is necessary to have a strong management system. One of the most appreciated is the balanced scorecard, which is a strategic planning and management system. It has been created to add the strategic nonfinancial performances to the more traditional one, in order to give a more "balanced" view to the managers. The balanced scorecard provides guidelines on what should be addressed, and how to measure it, helping the planners to effectively apply their strategies. This management system helps companies to put their vision and strategy into action. At the core of the balanced scorecard system, there is the idea that an organization should always be analyzed from four different points of view: the financial, the internal business processes, the learning and growth, and the customer. For each one of these, metrics, data, and targets should be analyzed [17].

HOW METRICS OF SUSTAINABILITY CAN BE USED (E.G., ASSESSMENTS, AUDITS)

There are different ways in which this broad panorama of different metrics can be used. For instance, some of these guidelines and standards are often used to gain certification, which can easily be used to "boost" the green image of the company, or used mainly as marketing tools. However, the purpose of implementing measures is far from this; for instance, it would be preferable to see them as a tool to quantify and improve sustainability.

Assuming that the constructive approach is the one overcoming the others, it is interesting to consider how those metrics are used to reach sustainable goals. Once effective key performance indicators (KPI) are founded, metrics are an extremely useful tool, which can be used to assess activities within sustainability, analyze achievements, and set new goals.

A defined metric allows an organization to have precise parameters to consider tracking progress and applying tailored management practices. The most common way of doing so is through periodic audits within the organization, as well as the supply chain. Annual assessment, or auditing, is really a powerful tool to gain a deep understanding of the corporate sustainability status including sustainability of supply chain (e.g., products). Ideally a third party carries out the process, since, clearly, this approach is much more fitting to a transparent and trustworthy behavior, which ordinarily characterizes an audit process.

Additionally the results given by the assessment or audit phase of the metric, if its parameters are clear and well explained, can be used to communicate more effectively the achievements of the organization. They can also be used to present an organization's best practices. Furthermore, most of the time the standards required are above the requirements established by the national law; thus, the compliance to such high standards will allow the organization to promote itself as a forerunner and a promoter of best initiatives [18]. As a consequence of this, it is important to choose the right label or right guideline, ones that have credibility.

Sustainability assessment or audits of organization's policies and its products is a holistic approach with many benefits such as:

- Selection of actions in order to effectively address the stakeholder needs
- Proper allocation of resources for the implementation of the selected actions
- Identification of opportunities for every stakeholder group
- Use of a comprehensive tool for the evaluation of strong points and areas for improvement
- Benchmarking of internal practices against selected guidelines, standards, or labels

STAKEHOLDER METRICS AND SUSTAINABILITY

There is obviously a strong correlation between key stakeholder perceptions and brand image, reputation of the organization.

For example, if customer satisfaction feedback shows a negative trend over a period of time, then there will definitely be a direct influence on market share, turnover, customer loyalty, and prices and therefore to sustainability. If employee's feedback is also negative over a period of time, then there is a direct influence on employee turnover, productivity, and many other areas that directly affect operational costs. If local community perception is negative over a period of time or after an ad hoc incident, then the image of the organization as a good citizen will be damaged and probably authorities, NGOs, will force pressure for change.

Additionally when the organizations implement systemic methodologies to understand and assess the key stakeholder perceptions, it means that they have comprehensive and accurate information but also the flexibility to react in an appropriate manner. If this information does not exist or is not systematic or is partial, then their ability to react based on real facts and in time is reduced significantly.

The following chart (Fig. 17.3) describes direct and indirect correlations between sustainability strategies and systematic stakeholder perception metrics supporting Sustainability business case:

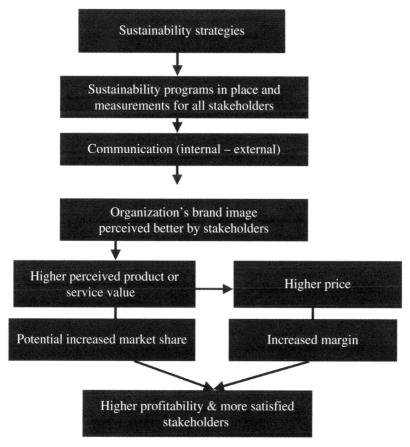

FIGURE 17.3   Direct and indirect correlations between sustainability strategies and systematic stakeholder perception metrics supporting sustainability business case.

REFERENCES

1. Alan F. Measuring up to sustainability. J Futures Stud 1998;30(4):367–375.
2. Global reporting initiative framework. Available at www.globalreporting.org/. Accessed 2013 Jul 30.
3. British Standard Institution – guidance for managing sustainable development. Available at http://shop.bsigroup.com/ProductDetail/?pid=000000000030118956. Accessed 2013 Jul 4.
4. International Standard Organization. Available at http://www.iso.org/iso/home.html. Accessed 2013 Jul 4.
5. ISO 26000. Available at http://shop.bsigroup.com/en/ProductDetail/?pid=000000000030140726. Accessed 2013 Jul 4.
6. Discovering ISO 26000. Available at http://www.iso.org/iso/discovering_iso_26000.pdf. Accessed 2013 Jul 4.
7. From the global footprint network – living planet report. Available at http://www.footprintnetwork.org/en/index.php/GFN/page/2010_living_planet_report/. Accessed 2013 Jul 4.
8. From the global footprint network. Available at http://www.footprintnetwork.org/en/index.php/GFN/page/footprint_basics_overview/. Accessed 2013 Jul 4.
9. The environmental footprint. Available at http://ec.europa.eu/environment/eussd/product_footprint.htm. Accessed 2013 Jul 4.
10. European Commission Environment, 2011. http:/ec.europa.eu/environment/index_en.htm. Accessed 2013 Jul 30.
11. From the International Organization for Standardization. Available at http://www.iso.org/iso/iso_14000_essentials. Accessed 2013 Jul 4.
12. ISO 14046 on Water footprint. Available at http://www.bsigroup.com/en-GB/ISO-14046-Water-footprint--Principles-requirements-and-guidelines/. Accessed 2013 Jul 30.
13. ISO 14067 on Carbon footprint. Available at http://www.iso.org/iso/iso_catalogue/catalogue_tc/catalogue_detail.htm?csnumber=59521. Accessed 2013 Jul 4.
14. Product carbon footprinting – a study on methodologies and initiatives by Ernst & Young. Available at http://www.saiplatform.org/uploads/Library/Ernst%20and%20Young%20Review.pdf. Accessed 2013 Jul 4.
15. From GHG protocol. Available at http://www.ghgprotocol.org/. Accessed 2013 Jul 4.
16. Available at http://www.quantis-intl.com/life_cycle_assessment.php. Accessed 2013 Jul 4.
17. The balanced scorecard concept. Available at http://www.balancedscorecard.org/BSCResources/AbouttheBalancedScorecard/tabid/55/Default.aspx. Accessed 2013 Jul 4.
18. Available at https://www.sustainablehighways.org/98/why-measure-sustainability.html. Accessed 2013 Jul 4.

CHAPTER 18

Life Cycle Analysis and Carbon Footprint

CLIMATE CHANGE AND CARBON FOOTPRINT

The rise of climate change on the international agenda has pushed companies and organizations to find different ways to assess, measure, and reduce their carbon footprint. In this chapter, we will present some concepts on this pressing issue, as well as how carbon footprint can be measured, and, finally, three relevant case studies from the business world.

Climate change has been defined by the International Panel for Climate Change (IPCC) as "...a change in the state of the climate that can be identified (e.g. using statistical tests) by changes in the mean and/or the variability of its properties, and that persists for an extended period, typically decades or longer. It refers to any change in climate over time, whether due to natural variability or as a result of human activity."

Numerous phenomena have been observed and registered, such as increasing temperature worldwide, raising of the oceans, melting ice, and increasing in numbers and intensity of the meteorological phenomena. Whether the climate change can be considered a cyclic event in time and modality that have already manifest in different geological eras, the current climate change seems to be linked to the high emission of greenhouse gases (GHG) produced by humans, a process that started since the industrial revolution at the end of the eighteenth century. Furthermore, in the document, the IPCC affirms that, for the second half of the twentieth century, "The observed patterns of warming, including greater warming over land than over the ocean, and their changes over time, are simulated only by models that include anthropogenic forcing."

Practical Sustainability Strategies: How to Gain a Competitive Advantage,
First Edition. Nikos Avlonas and George P. Nassos.
© 2014 John Wiley & Sons, Inc. Published 2014 by John Wiley & Sons, Inc.

To make projections of the possible trends of climate change, the IPCC produced the Special Report on Emissions Scenarios [1]. According to this study, the temperature in the last decades rose by 0.2 °C per decade, and according to different projections, within 2100 the surface temperature can rise between 1.8 and 4.0 °C. In a similar scenario the consequences, which the IPCC tried to define, concern the ecological resilience to such significant stress; the global reduction of food production, which will occur after a temperature increase above 3 °C; the rise of the sea level, which will increase erosion, floods, and migration flows; the potential enlarging of transmitting diseases, such as malaria; and, probably one of the most important points, the alteration of water cycle and consequently of the availability of water and its distribution [2].

CARBON EMISSIONS AND CARBON FOOTPRINT

As underlined in the previous chapter, the necessity of measuring carbon emissions has constantly and significantly risen in the last years. Probably one of the most significant inputs that pushed towards the reduction, and therefore quantification, of carbon emission was the Kyoto Protocol. The United Nations Framework Convention on Climate Change released this international amendment in 1997, and it entered into force in 2005, for the reached quorum of 55% of countries ratifying it [3].

The most common way to quantify carbon emissions is the carbon footprint. This footprint is a tool to measure the total amount of GHG emissions caused directly or indirectly by a company or a specific product. It is usually expressed in carbon dioxide equivalent (CO_2e); the reason for this is because CO_2 is not the only gas that is considered in this calculation, but, according to the Kyoto Protocol, calculations also include methane (CH_4), nitrous oxide (N_2O), hydrofluorocarbons (HFCs), perfluorocarbons (PFCs), and sulfur hexafluoride (SF_6). All these six gases, to facilitate the communication, are transformed into the equivalents of CO_2 that would cause the same warming effect [4].

There are different kinds of footprint that can be calculated according to the interest, such as individual, organization, or national footprint; however, the one most usually undertaken is that of companies.

The carbon footprint of organizations is usually divided into organizational and product carbon footprint. The first, which is easier to calculate and therefore the most adopted, contains all the emissions that are related to the energy consumption of directly owned buildings and sometimes also to waste production and transportation [5].

Product carbon footprint is more complex; indeed, it includes the emissions of the whole production chain of the product, going from the extraction of the raw materials to the final disposal (sometimes instead of the cradle to grave, the cradle-to-gate approach is adopted). As it is easy to imagine, this calculation is far more time and money consuming; however, it is also more precise [6].

Once the carbon emissions are quantified, the next step is to try to reach the neutralization or balance of the total emissions produced. Thus, the concept of carbon balance or offset was created. This idea pushes the organization, once it has undertaken all the initiatives possible, to reduce its CO_2 emissions and to balance its remaining ones. In order to do so, a company has to develop projects to reduce or storage CO_2; some examples of these may be tree planting, creation of a wind farm, installing solar panels, or carbon storage [7].

However, one of the biggest challenges related to climate change and carbon emissions is how to measure the carbon footprint. Indeed, in the last decade there has been a flourishing of different methodologies trying to achieve this. One of the most recognized methodologies is the GHG Protocol based on a Life Cycle Analysis (LCA) approach.

LCA AND MEASURING CARBON FOOTPRINT

The most adopted methodology for measuring the carbon footprint is LCA. It is a methodology for assessing the aspects associated with a product over its life cycle. The purpose of this study is to identify and improve the most pressing issues along the process of production (which typically are environmental but can be also social and economic). To do so, the stages, which are usually considered, from cradle to grave, are extraction/production of raw materials, manufacturing, packaging and distribution, product use, and final disposal.

The International Organization for Standardization (ISO) has provided guidelines on how an LCA should be conducted, and it did so in the ISO 14040 and 14044, describing the four main stages:

- Goal and scope definition, where the product and the methodology of assessment are defined
- Inventory analysis, where all the inputs and outputs of extractions and emissions have to be listed
- Impact assessment, which defines impacts related to these inputs and outputs
- Interpretation, where the results are reported informally and related to the scope and goal of the study [8]

GHG PROTOCOL

The GHG Protocol is one of the most used standards providing a methodology for calculating the carbon footprint. It was developed by the World Resources Institute, in collaboration with the World Business Council for Sustainable Development, and released in October 2010. The GHG Protocol comprises four different, but still related, standards: the Corporate Accounting and Reporting Standards (Corporate Standard), the Project Accounting Protocol and Guidelines, Corporate Value Chain (Scope 3) Accounting and Reporting Standard, and the Product Life Cycle Accounting and Reporting Standard.

The Corporate Standard is a methodology that provides a method to report on the total GHG emissions of the organization. The Project Protocol gives guidelines on how to quantify the improvements achieved by reduction activities (there are specific protocols for Land Use, Land-Use Change, and Forestry (LULUCF) and Grid-Connected Electricity Projects). The Corporate Value Chain (Scope 3) Standard provides the tool to quantify the total emissions of the supply chain (which usually are the most significant ones) and how to reduce them; here, there are usually included also suggestions on how to face the issues with partners and suppliers. Finally, the Product Standard is used to have a complete overview of the emissions over the entire life cycle of a product (raw materials, manufacturing, transportation, storage, use, and disposal) and identify the biggest reduction chances, in terms of emissions, efficiencies, reducing costs, and removing risks; the standard also helps in providing communication [9].

The GHG Protocol has been successfully implemented worldwide because it can be easily applied to any kind of product or service and is based on LCA. This theoretically makes it suitable to be applied on other environmental impacts (even if the focus stays on GHG). Furthermore, the standard helps companies to identify emission reduction opportunities and risks in the supply chain, making improvements over time; communicate; engage partners; etc. [10].

LCA AND PRODUCT LABELS

The LCA also has applications other than being used by international, or national, institutions to release guidelines (such as the GHG Protocol). Indeed, it is probably the most adopted methodology also for the development of standards and labels. The LCA of a product has been positively recognized as a powerful tool for companies in order to understand their environmental impact and costs throughout the supply chain and to succeed in reducing them.

The LCA of the carbon footprint of a product usually takes into consideration the emissions of carbon dioxide during the whole life cycle, which, if not stated differently, is extraction/production of raw materials, manufacturing, packaging and distribution, product use, and final disposal. The final value is generally presented in CO_2e in kilograms, pounds, or metric tons.

Carbon footprint can be used by business as an environmental indicator. To do so, however, organizations have to find the right balance between the complexity of such analysis and the difficulty of communicating it effectively to the customers (here is where the labels should help).

The Carbon Disclosure Project (CDP) was one of the first to develop a reporting tool questionnaire for the supply chain, asking companies to look deeper in their emissions of GHG and report them, as well as the indirect emissions. It is interesting that, since 2009, Wal-Mart tries, together with its enormous list of suppliers, to increase quality and decrease costs. This aim should be attained by the reduction of costs related to manufacturing, energy use, carbon emissions, waste, and packaging and engaging the suppliers in the improvements. And, indeed, this is exactly what characterizes the initiative of Wal-Mart: the strong relation with suppliers and partners, which is actually helping in spreading the costs and boosting the brand recognition [11].

On the other side, the communication towards the final customer is a complex field, and here labels can have a key role in facilitating the process. The first, and probably one of the most adopted labels, is the one released by the Carbon Trust together with the UK Department for Environment, Food and Rural Affairs (Defra) and BSI British Standards, the Carbon Reduction Label. This label comprises a black footprint in which are written the values in grams or kilograms per tons of CO_2e. In order for an organization to use this label, it should calculate the exact carbon footprint of its products through the PAS 2050 Standards [12]. But this label is not the only one; indeed, in the last years there has been a flourishing of different labels also outside the United Kingdom. Some of these are *LÍndice Carbone* adopted by Casino, and developed by BioIntelligence; the Greenext Index developed for E. Leclerc; the *climatop* carbon label used by Migros (Switzerland); and the Certified Carbon Free in the United States, released by the Carbon Fund in collaboration with the Edinburgh Center for Carbon Management on the basis of ISO LCA standards, the GHG Protocol, and the UK Carbon Trust's (2007) Carbon Footprint Measurement Methodology; the German government released a "Product Carbon Footprint." The China Energy Conservation Investment Corporation (CECIC) in collaboration with the Carbon Trust is creating one, while the European Union has commissioned to an Italian consulting, Life Cycle Engineering (LCE), to develop a Carbon Footprint Measurement Toolkit in cooperation with the Swedish Environmental Management Council (MSR); and the list can go on [13].

However, it is interesting to notice that the first retailers to develop a program of labeling the carbon emissions of products, Tesco, after labeling 500 products and doing research on 1100, were also the first to abandon this path—after four years. The reasons that Tesco is providing for this decision are numerous. Firstly, the costs and time for each analysis on the product are high; indeed, each good involved "a minimum of several months' work." Secondly, Tesco was expecting that other retailers would also have moved in this direction fast, but they did not [14].

Carbon labeling can be seen as a positive chance for organizations that should, however, approach this complex issue with a cross border, sector, and phase view, keeping the balance between the precision and useful side of their path towards carbon labeling [15].

CASES FOR CARBON FOOTPRINT

Aviva, which is the fifth biggest insurance company worldwide (based in the United Kingdom), is working mainly on long-term savings, fund management, and general insurance. Its aim is to work to reduce its carbon emissions and offset all the residual emissions, to become the first insurance company to neutralize its carbon emissions related to its worldwide operations. The decision to go carbon neutral comes in 2006 and rises from desires of the group to:

- Be responsible and minimize its contribution to causing climate change and encourage others to do the same
- Maintain its forward-thinking reputation and create a market differentiator
- Create competitive advantage to strengthen opportunities to access new markets (e.g., socially responsible investment) or new products with environmental focus (e.g., discounts on hybrid cars)
- Increase employee engagement and trust by motivating staff to take action to reduce energy use and improve energy efficiency in the office and, in turn, at home
- Identify business risks associated with climate change by taking a proactive approach

Before that announcement Aviva had already reduced its carbon emissions related to its buildings and travels by 54%. In order to reach this ambitious goal, Aviva puts the carbon management as part of the corporate objective and is working on different aspects, which are energy consumption and saving, waste management, paper use, and traveling; then the offset is contemplated.

For the energy efficiency they are focusing on high efficiency in a new building (with low-energy lighting, recyclable furniture and carpets, and plenty of natural light), while for the consumption already 55% of their energy comes from renewable energy, and they are encouraging behaviors that favor the reduction of consumption. They plan to increase the recycling rate for waste, already at 59%, and they have replaced 400 printers with others that are more energy-efficient and paper saving. Last but not least, Aviva tries to reduce its travel emissions by promoting teleconference within the United Kingdom and with the adoption of an in-house car pooling.

For the final offsetting the company has been audited by a third party in 25 countries, which has evaluated the carbon equivalent that the company should offset. Then the projects about whom to offset with have been chosen internally—they are six and they have global focus and are a blend of social and commercial projects [16].

Dell is one of the world's biggest computer manufacturers in the world. They claim to be the first in the sector to commit themselves to neutralize their carbon emissions. Dell's commitment to neutrality concerns mainly its emissions coming from electricity use, heating, and cooling. Since 2003 Dell has reported data of their emissions to the CDP, and since 2007 they have joined the CDP's Supply Chain Leadership Collaboration Project to report the emissions of suppliers and help them to develop addressed strategies, in order to reduce Dell's indirect climate impacts. In 2010 Dell started to report the carbon footprint of one of its products, the Latitude E6400 laptop, and reduce the environmental impact of its products (together with suppliers). To reach its goal of 15% reduction by 2012, Dell is making the inventory of its organizational emissions and pushing for high-energy efficiency and with a high use of renewable energy (and offset the rest); furthermore, it also commits to design more efficient products. It has to be mentioned that, in the last Corporate Responsibility Report, Dell is claiming that it has reviewed the way of measuring and setting goals on carbon emissions, without specifying how. Lastly, they also say that they will abandon the Renewable Energy Credit program and they will use renewable energy just as much "as practical" [17, 18].

Barclays, which is one of the biggest banking and financial groups worldwide, is trying to minimize its direct impact of its operations through the adoption of a group-wide environmental management system. Then the remaining emissions are offset buying carbon credits. To do so, Barclays has different offset projects around the world; in 2010 it purchased 1,191,956 carbon credits, from projects in Brazil, China, India, South Korea, Tanzania, Kenya, and Thailand, to offset their energy and travel carbon emissions.

The project that has been undertaken in Brazil addressed over 1000 hectares, which were previously run as extensive grazing for cattle, where more than 50 endemic species have been planted. The adoption of different species guarantees the reconstruction of a more "natural" ecosystem, which facilitates the restoration of biodiversity in the area. The offsetting due to this project in Brazil was of 15,000 carbon credits. While the carbon storage was the primary purpose of this project, a wide variety of other benefits can be pointed out within the environmental and social sphere. Indeed, the tree nursery, together with the monitoring and maintenance of the forest, requires that people from the local community are hired and competently trained. From an environmental point of view, this project will help the restoring of a precious ecosystem such as the Brazilian rainforest—which every year loses approximately between 30,000 and 40,000 square kilometers.

Case Study: Eco Labels Bases on LCA—*climatop*

Using LCA products carbon footprinted from production through to disposal, carbon footprint is undertaken and audited by an independent third-party organization to ISO14040 standards.

Products can display the *climatop* label if their carbon footprint is relatively (at least 20%) lower than ones of a representative reference group, as well as complying with a set of environmental, social, and ethical requirements. Products are allowed to display the label for two years before they are reassessed.

Market Coverage   There are currently 60 products using the *climatop* label. Full details of the companies are available through the *climatop* website (www.climatop.ch).

WORLD POPULATION AND ENERGY CONSUMPTION

Today seven billion people live on our planet. Eighty percent of all human beings are able to cover their basic needs for food, clothes, living space, medical care, and energy for mobility, heat, and coldness. In comparison, at the time of the French Revolution in 1789, almost one billion people lived on our planet and most of them in extreme poverty. These successes have been made possible thanks to the spectacular advances in the field of science and technology and thanks to the continually growing use of raw materials, resources, and energy. But environmental pollution as well as dwindling fuel and energy resources are the other side of the same coin.

Many problems, such as environmental pollution or the use of raw materials, can be solved. The high-energy consumption and the climate warming have the potential of being disastrous for the world.

ENERGY CONSUMPTION AND GLOBAL WARMING

Energy is quite special. The availability of an adequate amount of energy is necessary for every aspect of our life. The production of food and the recycling of materials are just two out of many examples. Nothing runs without it. Unfortunately, the climate change is strongly linked to the wastage of energy.

In the near future, our planet will be inhabited by more than nine billion people. The high rate of population growth as well as the substantial rise in prosperity in newly industrialized countries leads to a strong demand for energy and raw materials. Due to improved prosperity, there is a higher demand for meat and dairy products. They show a significantly poorer climate balance than cereals. The global problems caused by climate change and the lack of energy sources will continue to grow.

climatop's Contribution

We are convinced that the humanity is able to solve their problems concerning energy supply and climate change. We contribute a part to the solution in form of the *climatop* label.

The energy and climate issues may be solved thanks to innovations, which are attained by higher prices/costs. Two opinions should be proceeding parallel:

- Saving energy
- An increased and additional generation of renewable energy

According to all results of calculations, it is cheaper to save energy instead of producing new one. Nonetheless, saving alone won't be sufficient, which is why we need new alternative energy sources. An essential energy-saving aspect is the use of innovative products, like the Dyson Airblade™ hand dryers, which need less energy for the same usage/function.

At that point, *climatop* may come into play. *climatop* calculates and compares innovative products and certifies those who need, compared with a representative peer group or an industry average, less energy or, in the case of agricultural products, emit less green house gases. Moreover, only products with equally good total environmental impacts are labeled.

REFERENCES

1. SRES, 2000. Available at https://www.ipcc.ch/pdf/special-reports/spm/sres-en.pdf. Accessed 2013 Jul 31.
2. IPCC Climate Change 2007. Synthesis report. Available at http://www.ipcc.ch/pdf/assessment-report/ar4/syr/ar4_syr.pdf. Accessed 2013 Jul 4.

3. From the United Nation framework convention on climate change. Available at http://unfccc.int/kyoto_protocol/items/2830.php. Accessed 2013 Jul 4.

4. The carbon trust and carbon footprint. Available at http://www.carbontrust.com/client-services/footprinting/footprint-measurement. Accessed 2013 Jul 4.

5. Organizational carbon footprint. Available at http://www.carbontrust.com/client-services/footprinting/footprint-measurement. Accessed 2013 Jul 4.

6. Product carbon footprint. Available at http://www.carbontrust.com/client-services/footprinting/footprint-measurement. Accessed 2013 Jul 4.

7. Carbon neutrality concept. Available at http://www.coco2.org/about_whatisit.php. Accessed 2013 Jul 4.

8. LCA methodology, explained by PRe'. Available at http://www.pre-sustainability.com/content/lca-methodology. Accessed 2013 Jul 4.

9. The GHG protocol. Available at http://www.ghgprotocol.org/standards. Accessed 2013 Jul 4.

10. Ernst & Young and Quantis. Product carbon footprinting – a study on methodologies and initiatives; 2010. Available at http://wko.at/tirol/industrie/indakt2010/Folge38/PCF-report.pdf. Accessed 2013 Jul 4.

11. Kral C, Huisenga M, Lockwood D. Product carbon footprinting: improving environmental performance and manufacturing efficiency. Available at http://www.wspenvironmental.com/locations/product-carbon-footprinting/en. Accessed 2013 Jul 4.

12. The carbon reduction label. Available at http://www.carbontrustcertification.com/cust_maint/default/site_down/maintenance.html. Accessed 2013 Jul 4.

13. List of carbon labels. Available at http://www.tescoplc.com/assets/files/cms/Tesco_Product_Carbon_Footprints_Summary(1).pdf. Accessed 2013 Jul 31.

14. Tesco quit carbon footprinting. Available at http://www.guardian.co.uk/environment/2012/jan/30/tesco-drops-carbon-labelling. Accessed 2013 Jul 31.

15. 'Following the footprints': Technology quarterly, *The Economist*, Q2 2011. Available at http://www.economist.com/node/18750670. Accessed 2013 Jul 4.

16. From the study conducted by Article13. Available at http://www.article13.com/CBI/CBI_CSR_Case_Study_Aviva_Sept%2007.pdf. Accessed 2013 Jul 4.

17. Dell corporate responsibility report. Available at http://i.dell.com/sites/content/corporate/corp-comm/en/Documents/dell-fy11-cr-report.pdf. Accessed 2013 Jul 4.

18. Available at www.cleanair-coolplanet.org. Accessed 2013 Jul 31.

CHAPTER 19

Water Footprint

OVERVIEW

A.Y. Hoekstra introduced the water footprint concept in 2002; it has been developed as an indicator, used to express the total amount of volume of freshwater consumed to produce a particular good or to provide a service.

The main objective of assessing the water footprint of goods is to analyze how their production is affecting water resources (and pollution) and how these processes can be more sustainable (from a water perspective).

The water footprint is composed of three components: *green*, *blue*, and *grey water footprint*. The *blue water* represents the volume of water, which is withdrawn from surface (sea, lake, and river) or groundwater, and does not return to the same basin of catchment, or to the sea.

The *green water* is the rainwater, the soil moisture adsorbed, and the evapotranspiration of the plant. As it is easy to imagine, this component of water footprint is of particular relevance towards the agriculture sector in comparison to the industry sector (just in the case in which they have green roof, but, even in that case, is usually not significant).

The last component of the water footprint, the *grey water*, represents the amount of water, which is necessary to dilute the load of pollutants/emitted substance from the production chain, in order to make them harmless.

The water footprint can be expressed in a variety of ways, according to what is most relevant to underline in the assessment. For instance, it can be expressed as water volume per piece (when the product is countable) or as water volume per product unit (mass, money, piece, unit of energy, etc.). However, the classical way of its representation is in the form of water volume per unit of time (cubic meter per ton or liters per kilogram).

The water footprint can be calculated on different scales, starting from the individual consumer to companies, products, and even entire countries. Such

Practical Sustainability Strategies: How to Gain a Competitive Advantage,
First Edition. Nikos Avlonas and George P. Nassos.
© 2014 John Wiley & Sons, Inc. Published 2014 by John Wiley & Sons, Inc.

categories are usually divided into direct and indirect water footprints; for example, an individual is composed of the water consumed at home (direct) and the one comprised from the consumption of products and services (indirect). As for companies, the operational represents the direct consumption, while the suppliers/production chain the indirect [1].

The last concept that is worth mentioning about water footprint is that of water neutrality.

This concept is relatively new and was defined for the first time in 2007 by a heterogeneous group of six organizations. The group defined three main criteria characterizing this idea, being:

1. Defining, measuring, and reporting one's "water footprint" (*this is not always considered mandatory, especially for actors different from organizations*)
2. Taking all actions that are "reasonably possible" to reduce the existing operational water footprint
3. Reconciling the residual water footprint (amount remaining after a company does as much as possible to reduce the footprint) by making a "reasonable investment" in establishing or supporting projects that focus on the sustainable and equitable use of water

Unless these points have still some unclear aspects—such as what is reasonable and how to measure improvement and accounting for loss of water in different areas—the water neutrality concept can still be considered a useful tool to decrease the overall water consumption and increase the efficiency of use within the production chain [2].

GUIDELINES FOR MEASURING THE WATER FOOTPRINT

At the contrary of what is the situation for the calculation of the carbon footprint, for which there are plenty of different methodologies to use, for the water footprint, the choice is significantly smaller. Indeed, the most used framework is the one provided by the *Water Footprint Network*, which gives guidelines on how to assess the water footprint of a product. According to it, a complete water footprint assessment is made of four different phases:

1. *Setting the goals and scope: in order to define what is to be measured (consumption of water for the supply chain, one stage of production, final consumer, etc.)* The first thing to determine is which type of water footprint to calculate (product, consumer, production step, etc.) and to what level of detail. Then the boundaries should be set, defining

what is included in the study and what is not, for example, which component of the water footprint (green, blue, grey), spatial–temporal borders, direct or indirect consumption. The sustainability assessment, which is usually based on geographic perspective, enables the analyst to link the hot spot and issues to a particular place or reality. Finally, the response formulation here is particularly important to individuate who should be involved and participate to the "solution" process started by the water footprint analysis.

2. *Water footprint accounting: here the water consumption is quantified and located in space and time*   Its aim is to measure the human appropriation of freshwater in a determinate catchment area—expressed in volume. The water, which composes the water footprint typically, originates from the one catch by evapotranspiration and runoffs. The water footprint of a single process-step is the basic measure for every accounting of water footprint; then the intermediate or final product calculation can also be done and add to the total accounting. The accounting of water footprint (which is "the water footprint of a product is defined as the total volume of freshwater used directly or indirectly to produce the product") is therefore calculated by including the consumption and contamination of water in all the stages of the production chain (this process will almost always be the same, no matter the sector).

3. *Water footprint sustainability assessment: is where the environmental, social, and economic aspects are evaluated*   The suggestion given by the guideline considers the assessment of the sustainability process of the catchment of water according predominately by two parameters: the location in time and space within a catchment or river basin and from which kind of process or group it is addressed (i.e., producer or consumer). For the calculation of the sustainability of the water collected, different impacts within environmental, social, and economic spheres are taken into consideration (once relevant criteria are identified). Then precise suggestions on how to calculate the sustainability of processes, products, and consumers are given.

4. *Water footprint response formulation: in which a strategy is trying to be developed*   The guideline is intended to be a useful tool to understand in which direction solutions should come and go. The main actors that are addressed are consumers, companies, farmers, investors, and governments, and for each of them, suggestions are given on how they can act and undertake initiatives to reduce the consumption of water and therefore the water footprint. Lastly, there is also a short mentioning on how it should be preferable to have clear targets of reduction instead of an offsetting program to reach water neutrality [1].

Another interesting tool that is under development and was released within 2012 is the ISO 14046 or Water footprint—Requirements and guidelines, developed by the International Organization for Standardization (ISO). The work of ISO is aimed at integrating the standards of Life Cycle Assessment (LCA) and the carbon footprint standard ISO 14067 (which is also under development by ISO). The purpose of this guideline is to include:

- Principles and guidelines for a water footprint metric of products or organizations
- How different types of water (green, blue, and grey, as well as ground and surface water) should be considered together with the socioeconomic and environmental issues
- How to communicate on water footprint
- How to make it compatible with the other standards of the ISO 14000 family [3]

WATER FOOTPRINT AND LCA

At the moment the most commonly used framework to calculate water footprint is the guideline released by the Water Footprint Network, which is assessed mainly through water inventory and accounting.

Indeed, water footprint is sometimes criticized for the absence of "characterization factors," which should weight its components according to their impact, and will be appreciated for a LCA methodology (and perspective). But this, according to the Water Footprint Network, will have an influence on social and environmental impacts, since it will lead to the omission of important key indicators—such as the variability in time and space.

To agree on both the purposes of Water Resources Management and LCA, the steps usually undertaken for the calculation of the water footprint should be approached differently in a LCA perspective. To do so, the accounting stage will be part of the inventory of LCA, while the impact assessment and aggregated impact assessment will be part of the life cycle impact assessment. But this, according to the authors, will still invalidate the effectiveness of the water footprint meaning [4].

CASE STUDY: SABMILLER

SABMiller is the second biggest brewing and bottling company in the world and has several programs running to improve its sustainability. Particularly interesting are its initiatives for the reduction of water use and how it is reporting on it. Indeed, hereafter will be presented some case studies among some of its production sites, which are taken by its Water Footprint Report, written with WWF-UK, and released in 2009. In this report SABMiller presents two case studies of its subsidiaries, one in South Africa and the other in Czech Republic. At first they undertake a study on the average availability of water within the two countries of analysis. In South Africa they observed how their plants and distribution network are spread on the territory, and due to this situation (as well as others), they decided not to include the grey water footprint in the study. The biggest component of water footprint is due to the green water component, which is 98.3% due to its crop consumption (locally 84.2% while the imported one is just 14.1%). However, a more geographically detailed analysis has pointed out how the consumption of the green and blue water changes significantly within the country according to the weather/climate conditions. Indeed, in the south part, the crops were completely relying on precipitation, while the north crops are growing with a contribution up to 90% of irrigation. The collected data are then related with the potential impact of the territory and the local context; this showed that actually the rain-fed crops were more vulnerable to the availability of water on the long term (because of climate change and population growth), instead of the one relying on irrigation, which has more reliable source of water for the longest period. Lastly, SABMiller has also undertaken a research on the cost of the water within the value chain discovering that most of the water costs related were coming from treatment of their gray water by the municipality.

Instead, the picture and the reality inquired in the Czech Republic have revealed a different situation (selected because it produces 20% of European volume of beer). In this study, compared to the one in South Africa, the data was collected during a three-year period (2006–2008), therefore giving more strength to the methodology. Also here most of the water consumption is given by the crop's green water (around 90%, of which 75% is grown locally and 24% from the imported one); the irrigation represents just 6% of the water footprint. This longer period of collecting data allowed seeing how the irrigation mode is influenced by the weather and the precipitations and therefore the composition of the water footprint. What is interesting to underline as well as emphasizing that the report is doing as well is that both studies are easily comparable, even if from different realities (geographically) thanks to the standardized methodology of measuring the water footprint—the guidelines of the Water Footprint Network [5].

NESTLE

Nestle has developed a drip irrigation project in Nicaragua, to develop a low-cost irrigation system to use in the coffee plantations. This project took place during a three-year period (2007–2010), and Nestle developed it in collaboration with ECOM, Rainforest Alliance, and IDE, and it estimated the participation of 1500 small coffee producers.

The purpose of the project was to provide the coffee crops with a supplementary irrigation system during water stress periods (and flowering periods), since this has shown to increase the productivity, the growth of the plant, and the quality of the crop. Indeed, during the critical growth period, a lack of irrigation could have a strong negative impact on coffee performance and quality. To do so, Nestle introduced, improved, and recorded a low-cost dripping system, as part of a process of embedding sustainability within its value chain.

The project started with the installation of a drip irrigation system in 11 different plantations in Nicaragua. The increase of production was registered the same year, for an amount estimated between 40% and 60% of the yield of the previous year, and at the same time young plants grew faster. These results were confirmed during the following years of the projects, which also allowed observing that irrigated plants of two years were producing similarly to plants that had not been irrigated over a period of three years.

Since the successful results of this project, the system of irrigation has been implemented in other plantations in Nicaragua (32 in total) and also in Honduras (9 systems), El Salvador, and Guatemala [6].

The *Dole Food Company*, which is the world's biggest producer of fruit and vegetables, has introduced a new method to reduce the use of water consumed during the packing process of bananas. To allow a sensitive reduction of water consumption, Dole has implemented different types of water recycling systems during the packing process in the last two decades.

The use of water in the packing stage takes place mainly in three different phases: to remove dirt and insects from bananas, hold and carry bananas while they are selected, and for the removal stage. This traditional way of packing bananas requires around 150 liters per box (18.14 kilograms).

A first measure that Dole undertook at the beginning of the 1990s was the adoption of sand and other gravel filters inside a partial recirculation system in order to reduce the amount of water per produced box to 100 liters. In addition, a Mobile Banana Processor was also developed that allowed a 97% reduction of water use, which is distributed in remote areas of the Philippines.

After a few years Dole installed 120 new systems that completely recycle and recirculate the water for the packing process in Ecuador, bringing the water consumption for banana boxes down to 18 liters. Afterwards, this system was applied in water-scarce areas in Honduras, Colombia, and Costa Rica.

The last improvement that Dole designed was part of the "New Millennium Packing Plant." In this new system, rather than using water pools (as is common practice), most of the activities take place in the banana crop, leading to a 90% decrease of water use (and 50% less energy) in comparison to a traditional system. Dole's specific initiative has been awarded with an "honorable mention" from the Scientific Committee of the World Water Week in Stockholm in September 2010 [7].

WATER FOOTPRINT VERSUS CARBON FOOTPRINT

The numerous "footprint" indicators that have been born in the last few years have been intended as complementary indicators of the environmental impact of human activities, since they focus on different issues.

Both water footprint and carbon footprint are two really useful tools for the quantification of the emission or consumption of the respective resources, as well as of the improvement that they can lead during their assessment. However, both footprints communicate just the final value of the study. This can be really deceptive (particularly for water), since the amount of water consumption communicated is not related to the hydrogeographical area; therefore, it is not necessarily synonymous with water stress in the area. Indeed, if we compare it to the carbon footprint, where we assume that any quantity of CO_2 emitted (and other GHG) is increasing the greenhouse effect (therefore it has a "negative" impact), we cannot say the same for the water consumption. In fact, a high consumption of water in a particular region/basin should not automatically lead to a water stress situation. As a consequence, it is easy to understand how it can be difficult to communicate effectively the assessment of the footprint undertaken.

Regardless of the communication problem, the water footprint and the carbon footprint remain really valuable tools for companies wishing to find their major point of consumption and waste, in order to quantify and try to reduce their water consumption (e.g., through the identification of hot spot). A challenge regarding the future use of different footprints, which is recognized by different experts, is the development of a framework that would be able to integrate all these different analyses in a comprehensive and useful manner.

The last point of difference between these two "footprints" can be found in their way of offsetting. Indeed, both of them have developed a system, which allowed offsetting and neutralizing the emissions or consumption through projects, in order to reach the "neutrality" balance. Here the only difference among the two strategies of offsetting is that the one for water has not yet developed an exchange system of quote among nations or companies since water is a local resource.

REFERENCES

1. Hoekstra AY, Chapagain AK, Aldaya MM, Mekonnen MM. *The Water Footprint Assessment Manual – Setting the Global Standard*. The World Bank Group; 2011.
2. Water neutrality. Available at http://www.bsr.org/reports/Coke_Water_Study_March_2008.pdf. Accessed 2013 Jul 4.
3. ISO 14046 on Water footprint. Available at http://www.bsigroup.com/en-GB/ISO-14046-Water-footprint--Principles-requirements-and-guidelines/. Accessed 2013 Jul 31.
4. Water footprint accounting, impact assessment, and life-cycle assessment. Available at http://www.waterfootprint.org/Reports/Hoekstra-et-al-2009-Water Footprint-LCA.pdf. Accessed 2013 Jul 4.
5. SABMiller water footprint report. Available at http://www.sabmiller.com/files/reports/water_footprinting_report.pdf. Accessed 2013 Jul 4.
6. From Nestle. Available at http://www.nestle.com/csv/case-studies/AllCaseStudies/Pages/Drip-irrigation-project-Nicaragua.aspx. Accessed 2013 Jul 31.
7. Dole Food. Available at http://dolecrs.com/sustainability/water-management/water-recycling-programs-for-banana-packing/. Accessed 2013 Jul 4.

CHAPTER 20

Green Marketing, Communication, and Greenwashing

GREENWASHING

Often greenwashing is not an outright attempt to be deceptive, but it stems rather from failing to consider environmental impact measures with the same robust attention as is usually given to more established and familiar measures of business performance.

Greenwashing—making exaggerated environmental claims in order to curry consumer favor—is one of the banes of the sustainability community. When a company, lauded for environmental performance, is revealed to be engaging in environmentally dangerous practices, it provides skeptics with the fodder they need. Even when the claims are found to be inflated (or impossible to substantiate), credibility is lost. For a business model based on the paradigm that transparency leads to credibility, trust, and ultimately market advantage, these incidents can be devastating.

For some companies, the desire to appeal to "green" consumers and to be perceived as a good corporate citizen is enough to encourage deliberately deceptive claims. However, most companies do in fact embed "green" practices within day-to-day operations and initiatives; therefore, on average, the main reason for greenwashing has nothing to do with malfeasance or bad intent. They have to do with the fact that more often we see a fragmented approach to rolling out projects or activities—where one department may be donating towards reforestation, while another is measuring energy consumptions and verifying their carbon footprint. Both initiatives are green, but do they reflect the core business? Is there substantial reasoning behind why these two departments have taken on these specific activities, and why they

Practical Sustainability Strategies: How to Gain a Competitive Advantage,
First Edition. Nikos Avlonas and George P. Nassos.
© 2014 John Wiley & Sons, Inc. Published 2014 by John Wiley & Sons, Inc.

are completely independent of each other? How is this reasoning being justified and measured against the financial and nonfinancial performances of the organization?

The old adage in business—that which is measured gets done—implies the reverse as well, that things that are not measured (or are not measured well) are often prioritized lower. This can create a negative feedback loop, which certainly has too often been the case for sustainability.

GREEN MARKETING, COMMUNICATIONS, AND SUSTAINABILITY

What does this have to do with greenwashing? Everything. Because companies that would never consider having newer, temporary, or inexperienced people produce their quarterly financial reports are completely comfortable with having them produce sustainability reports (sustainability performance measures). When these numbers fall short of standing up to scrutiny or are used to solely generating marketing or public relations materials, organizational reputation is at risk. Failing to recognize the potential dangers of this trend in organizations not only places reputation at risk but also places a chain of other dependent factors—employees, community, shareholders, and partnering organizations—at risk as well.

If embedded effectively, by qualified *Sustainability Practitioners*, in collaboration with key decision-makers, sustainability offers solutions to facilitate greater trust in business by applying fundamental principles to core business values and operations that will generate clear indications of how credible these claims are. As a result, there is greater transparency between facts and claims, thus enabling consumers to see an alignment between the brand identity, product characteristics, and acclaimed "responsible"/"good" corporate citizenship initiatives.

The competitive advantage of sustainability is the quality of ROI that emerges as products and services take on green characteristics and communications unite the brand's core values with consumer passion, stakeholder awareness, and transparent annual reports. Value for money then does not remain stagnant upon the purchase by a consumer, but rather suspends its impacts to include all relationships formed around the brand—both financial and nonfinancial. Organizations that focus their vision on sustainability and have a clear strategy, which incorporates standards and models, will enable the adoption of best practices, accelerate performance and ensure transparent communications.

GREEN MARKETING

Green marketing is the marketing of products that are presumed to be environmentally safe or friendly. Thus green marketing must take into consideration a broad range of activities, including product modification, changes to the production process, packaging changes, as well as modifying advertising according to American Marketing Association.

Consumers worldwide are increasingly admitting that their behavior has changed in the last few years to benefit the environment mainly from their purchasing decisions. Therefore, stakeholders' purchasing decisions for products worldwide have been affected by green marketing messages. However, organizations should not get too comfortable with this trend in purchasing habits. Statistically, a consumer will spend on average 45 seconds reading a product label before making a "buy-not buy" decision! They are not passive buyers but rather information seekers and eager to align organizational claims on products with their expectations, knowledge, and perceptions of value for money. Their buying decisions extend beyond their encounters with green products to include access to information regarding environmental claims of the product from trustworthy sources apart from its producer in a timely manner. The breakdown of traditional media channels to all-inclusive, rapid, real-time, divergent commentaries and news releases also plays a key role in forming opinions on the credibility of these claims.

MATERIALITY AND SUSTAINABILITY

Organizations need to ensure the materiality of their claims. Materiality refers to constituents that are of high concern to stakeholders and of high strategic relevance to organizations.

Materiality is a very important part of a comprehensive sustainability strategy including green marketing. Overstatement of environmental attributes is a very common marketing behavior leading to greenwashing. An environmental marketing claim should not be presented in a manner that overstates the environmental attribute or benefit, expressly or by implication. Marketers should avoid implications of significant environmental benefits if the benefits are in fact negligible. For example, if a package is labeled "50% more recycled content than before" based on the increase in recycled content of its packaging from 2% recycled material to 3% recycled material, the claim, while technically accurate, is likely to convey the false impression that the advertiser has significantly increased the use of recycled material.

A comprehensive sustainability strategy is the very first step organizations should design in order to show that they "Walk the Talk." There are many organizations marketing their green products without having applied fundamental sustainability practices in their operations, thus increasing the risk of being accused as greenwashers. How *sustainability* is embedded within an organization influences the degree to which it is perceived as a good corporate citizen or green business.

GUIDELINES FOR GREEN MARKETING

In the United States, the Federal Trade Commission (FTC) [1], the government body that regulates and oversees marketing and advertising, has established general principles that contain specific guidance applicable to certain environmental marketing claims. Additionally, European Association of Communication Directors (EACD) [2] has released as well a guide for responsible communication.

One of the questions that is frequently raised about "green" products and services is whether or not there is a receptive public that justifies companies modifying their processes and procedures to capture what may be a small niche or transitional market. The very fact that the US FTC developed and released (for public comment) a set of new proposed green marketing guidelines provides an important affirmation that the government of the largest economy in the world believes in the long-term interest and importance of environmental claims in promoting goods and services. The fact that green marketing was seen as important enough to merit in sustainability eased attention indicates that the environmental impacts associated with goods and services are a long-term prospect.

That the FTC has asked for public comment also indicates that they are serious about making sure that the guidelines help consumers to make informed choices and that they want to make sure that stakeholders have an opportunity to have input into the process. An open, public process is critical to a program with both real and perceived value.

In general the guidelines focus on not only the importance for *accuracy* but also the need for *clarity*.

Accuracy

A recurring phrase that appears repeatedly and throughout the document is the need for claims to be based on "competent and reliable scientific evidence." This is defined as including "tests, analyses, research, or studies that have been conducted and evaluated in an objective manner by qualified persona and are generally accepted in the profession to yield accurate and reliable results. Such evidence should be sufficient in quality and quantity based on standards generally accepted in the relevant scientific fields, when consider in light of the entire body of relevant and reliable scientific evidence, to substantiate that each of the marketing claims is true."

This should provide a great deal of comfort for those who worry about the possibility of wasting time and effort focusing on overblown or exaggerated environmental concerns. Clearly the FTC is putting its weight behind the need for scientific consensus about what is being addressed. Hand in hand with this is the natural extension of this requirement—the need for robust and accurate measurement of environmental impacts of the products themselves, all along the supply chain from their sourcing, manufacturing, packaging, and distribution through to their use and ultimate disposal.

The guidelines take issue with and offer very little tolerance for deliberate "greenwashing" including not only misrepresentation but also the *omission* of salient facts. Stating that doing so is "deceptive if it is likely to mislead consumers acting reasonably," the guidelines clearly are looking at ensuring disclosure as well as offering guidance for claims.

Clarity

Using words like "reasonableness," the guidelines seek to establish the basis for a "reasonable consumer" standard. They make a credible attempt to cover all the bases by which consumers can be influenced and/or make decisions. The guidance covers "any environmental claims in labeling, advertising, promotional materials, and all other forms of marketing in any medium, whether asserted directly or by implication, through words, symbols, logos, depictions, product names or any other means."

Marketers will be held to a powerful standard; they "must ensure that all reasonable interpretations of their claims are truthful, not misleading and support by a reasonable basis before they make the claims."

In some cases, the guidance is very specific; for a product to be defined as recyclable, for example, the product not only must be recyclable; the infrastructure and means to recycle the product or packaging must be in place for 60% or more of the consumers within the markets where the products are sold.

Another area the guidelines cover is production. If the capture and reintroduction of excess materials is part of the standard production process (i.e., when scraps are routinely reintroduced into the production process), the product cannot take credit for being made with, or including, recycled materials.

The idea here is for those that distinguish themselves by going beyond the industry standard to be the only products that are able to gain the benefit from making the appropriate environmental claims.

The guidelines also seek to define the "rules of the game" when it comes to packaging, marketing, and advertising. Just as the government put an end to the practice of hiding clear glass balls in a bowl to give the illusion that a soup contained more meat and vegetables (by keeping them near the surface) in photographs and television commercials, the new guidelines set a framework for where and how products are positioned and even the use of the familiar triangle three-arrow recycle symbol—regulating that it be placed on the bottom on containers rather than near the name of the product when the container is what can be recycled rather than the product.

What is clear is that the FTC guidelines raise the bar in terms of expectations. As a result, companies may find that their current practices—even those that are industry leading—only go as far as to conform to the new guidelines, thereby setting a new standard. Leading companies may find that they must either extend their efforts or be willing to accept ceding their leadership position. Consumers may not look favorably on companies that were once seen as leaders when those same enterprises now find themselves engaging in behavior that meets the guidelines—and suddenly their exemplary efforts now are commonplace and common practice.

A recent study showed that physical measurements of carbon in the atmosphere are higher than they should be if all the environmental claims from all companies were added together. Clearly there is some inaccuracy in counting. The FTC guidelines seek to prevent double counting, by prohibiting companies from claiming and sharing environmental results. For example, having solar panels on your roof allows you to claim the *portion of your power generated*, but not if you sell the certificates. In that case, you have also sold the rights to characterize your power use as renewable. And if you have solar panels visible on your facility, unless 100% of the power consumed is generated, thus the company cannot claim "made with solar power" as this, while true, also implies 100% of the power used is from this source. Instead, the company must determine the proportion and is limited to making that assertion.

Truly leading companies will develop their own reporting that is more robust than the standards, perhaps using their framework and intent to guide their own actions and messages. A terrific example is the company that decided to apply safety standards not only to its own employees—as was required—but also to all contractors and subcontractors working on its job sites. The immediate result was an accident rate that seemed to "spike" upwards as they began counting all accidents and injuries, not just those from full-time employees who earned a paycheck with their company name on it. In response to the immediate questioning of what had suddenly gone wrong, senior management was in the enviable position of pointing out that the accident rate was still the same for full-time employees but that what was wrong was, in fact, an industry standard that did not value all people equally. This set a new standard and bar, above regulatory requirements, but by using the same measurement and metrics, the company was faithful to the regulations.

Exceeding and defining industry standards—or even business standards in general—is one way that forward thinking companies will leverage their efforts from the existing guidelines and do more. Doing so will establish that they are not "toeing the line" but rather "raising the bar." This can also help serve as a preventive (there is no point regulating a problem that does not exist) and help reduce costs. It also can serve as a marker for the company culture, facilitating human resources efforts to promote sustainability and retain top talent. By reducing risks associated not only with potential environmental damage but also with defending against complaints due to false marketing claims, companies that engage in proactive efforts also reduce costs due to potential fines for either.

SUSTAINABLE COMMUNICATIONS STRATEGY

Sustainability and communications symbolize two complementary areas of an organization's mode of operation. Sustainability represents a "holistic" management approach of re-evaluating the overall corporate strategy and operation, as well as monitoring performance via continuously assessing stakeholder engagement. It is an internal approach aiming at instilling sustainability into the overall management and operations of the organization.

Communications, however, represents a function, which aims at communicating and promoting the strategic direction of the overall organization, department, and operations. It represents an external approach to communicate to stakeholders the progress of the overall sustainability strategy. In addition, it aims at convincing stakeholders of the message's purpose, at validating the corporate objectives, and at obtaining support for engagement on sustainability issues affecting the organization's economic, social, and environmental impacts.

The main objective of this complementarity is the ability to transmit the right message, in the appropriate amount, through the suitable communication channels.

Sustainable communications is the process of foreseeing stakeholder expectations and, using a predetermined set of communication tools, being able to articulate the sustainability direction and objectives within the organization's business operations.

$$\text{Design} \leftrightarrow \text{Reinforce} \leftrightarrow \text{Manage}$$

Sustainable communications encompasses a new "breed" of actions within the field of communications that enable for a balanced and a transparent promotion of the organization's business and sustainability responsibilities. The process is composed of three phases, each with a respective strategy and all interdependent of each other.

DESIGNING THE SUSTAINABILITY COMMUNICATIONS STRATEGY

Regardless of the overall corporate objectives, defining a communications strategy shows attention to planning, an understanding of the situation, an ability to carry out the work, and clear identification of the goal.

CONDUCTING AN ASSESSMENT AND DEFINING THE CHALLENGES/ISSUES

Internal Assessment: Is the current communications strategy effectively communicating the corporate sustainability strategy and progress in a manner that will effectively achieve to integrate it within the working environment? Stakeholder mapping to determine the sustainability needs of internal stakeholders is the current communications strategy supported by employees. Does it motivate employee to get involved in organization sponsored initiatives and events?

External Assessment: This is external assessment of the organization's impact on the overall society and environment and determining how well the currently implemented communications strategy responds to the needs and expectations of external stakeholders and stakeholder mapping to determine the sustainability needs of external stakeholders. In this stage, assessment is conducted to determine the level of efficiency achieved using existing communication channels and external partners (e.g., NGOs, associations). Is there a policy behind partnering with external entities to achieve and implement relative initiatives?

Defining the challenges: Analyzing the results of the assessment and determining what are the current strengths of the current communications strategy and what are the areas that need to be improved.

DEVELOPING A SUSTAINABILITY COMMUNICATIONS STRATEGY

For each of the four sustainability pillars of the environment, society, Human Relations and marketplace, allocate critical internal and external stakeholder groups that need to be targeted in order to assure a sustainable communications approach. Using the results of the assessment, at this stage, the organization will already have a viewpoint of their needs and expectation; thus in this stage the objective is to design the necessary initiatives and actions that need to be enforced. This step needs to be accomplished in parallel with identifying the type of partnerships that need to be formed in order to design and implement the above initiatives and actions. Again, the results of the assessment regarding the effectiveness and efficiency of the corporate policies vis-à-vis external partners will work as resources in order to update the terms and conditions via which the organization will partner with these external partners. The objective here is to assure that all means of investment (i.e., monetary, HR) are allocated accordingly and respectfully of the overall corporate and sustainability strategy.

Defining the challenges: Setting targets and developing an action plan per stakeholder group and per initiative designed in order to monitor and assess performance.

IMPLEMENTING THE COMMUNICATIONS STRATEGY

The ability of the organization to effectively implement and sustain its communications strategy depends on the designed internal processes that enable for a continuous monitoring and management of the internal and external communications initiatives.

Implementing the communications strategy involves closely monitoring the formulated Key Performance Indicators (KPIs) (quantitative and qualitative) and being flexible enough to immediately respond in case an issue comes up.

Organizations are now "evaluated" for their philosophies and what they stand for. In our era of sustainable communications, what you do is more important than what you say!

This is one reason why in the past couple of years organizations have turned into cause-related and green marketing initiatives as a means to promote "business as usual" but in social and/or environmentally conscious way.

The portion of stakeholders (consumers, employees, NGOs, etc.) who care about social and environmental strategy is rapidly increasing and has a direct effect on purchasing patterns based on personal beliefs. Prior to making a purchase, these groups want to check the company's sustainability profile to ensure that certain social and environmental standards are respected.

By responding effectively to stakeholder expectations, the organization is able to communicate issues that matter, achieving active engagement, thus creating business value.

A critical element for the implementation of a successful sustainability communications strategy is the ability to engage with stakeholders in a transparent manner, using clear messages, in a joint dialog.

REINFORCING STAKEHOLDER INTERACTION

Once the communications strategy has been drafted and implemented, how can we ensure that it will appeal to stakeholders? Most importantly, how will their feedback be assessed in order to update the communications strategy?

There are innovative ways to interest stakeholders of course as per grasping the issues that do in fact matter to them.

The Importance of Social Media

Social media enables real conversation between the organization and its stakeholders. Organizations have the ability to inform stakeholders of news and events and be able to receive immediate feedback in a collective manner. Furthermore, social media provides the opportunity to stakeholders to engage with one another and be able to collectively provide their opinions and comments, via virtual "town meetings."

Furthermore, by using the platforms made possible by social media, the organization can provide incentives to stakeholders to access related Web links and reports that further promote the corporate message. Stakeholders can be directed to company websites and/or to online news distribution sources where information pertained to the organization is published and ready to be commented on.

The opportunities provided by social media enable organizations to actively interact with a wider stakeholder base by tapping into the limitless potential of the Web.

The Importance of the Company Website

The development of a comprehensive, user-friendly website has the ability to transmit messages in an optimum way.

Most importantly, if the website is designed keeping in mind the stakeholder groups (internal and external) impacting and impacted by the organization's operations, it can represent the primary platform for real, active, transparent, and bilateral communication.

The Web content should be developed to communicate as much information as needed. Providing too little information may leave stakeholder questions and concerns unanswered, exposing the organization to public criticism for lack of transparency.

Having assessed and determined the needs and expectations of stakeholders as they pertain to the sustainability context of operations, the organization can rest assured that it will be able to substantially communicate financial, environmental, and social initiatives.

Furthermore, the company website should be developed in such a manner that will transmit information in multiple ways:

1. Videos presenting the Sustainability strategy
2. A visual presentation of the sustainability report, which is a new trend
3. Interviews of executive members
4. Advertisements that demonstrate the sustainability communications strategy
5. Video documentaries of how initiatives evolved from a plan on paper to a real stakeholder success story
6. Testimonials by internal and external stakeholders
7. Videos presenting partnerships with third parties (NGOs, associations, etc.)

Sustainable communications strategy should reflect those issues that are at the core of organizations, the issues that are most important to the stakeholders. While the sustainability concept is rapidly applied to more organizations, stakeholders are becoming more aware and concerned about how the lack of sustainability influences them.

REFERENCES

1. FTC Green Guides: part 260 – guides for the use of environmental marketing claims. Available at http://www.ftc.gov/os/2012/10/greenguides.pdf. Accessed 2013 Jul 31.
2. Integrating and Communicating Corporate Responsibility, EACD Service Brochure; 2011.

PART 4

CONCLUSION

CHAPTER 21

Epilogue: Where Do We Go Now?

Human life has been on Earth for about 50,000 (or somewhere between 6000 and 200,000 depending on the source) years. If we assume that 50,000 is correct, for the first 49,800 or so years, man has lived in the open fields; in the mountains; in the forests; next to lakes, rivers, and oceans; or in the savannahs. This is called biophilia from the Greek word meaning "love of nature." As a result, this kind of life has been imbedded in the human gene as being normal. In the past 100–200 years, however, man has lived in cities consisting of brick, concrete, steel, glass, etc. Genealogically, this is not normal. Consequently, when a person wants to feel good, he or she goes to the ocean, lake, mountains, or open fields. That is why most people have paintings of landscape in their homes, and that is why most pictures in hospitals are that of landscapes. This makes people feel good. Is there a new norm?

Even many of the world's religions are greatly concerned about protecting the environment. For example, the beginning of the Sunday worship service in the Eastern Orthodox Church starts with "For favorable weather, an abundance of the fruits of the earth, and for peaceful times, let us pray to the Lord." Disrupting nature's life cycle invariably will cause some harm to the environment, something that is opposed by God. Patriarch Bartholomew, spiritual leader of the world's Eastern Orthodox Church, has verbalized this thought when he said: "If human beings were to treat one another's personal property the way they treat their environment, we would view that behavior as anti-social and illegal. We would impose judicial measures necessary to restore wrongly appropriated personal possessions. It is, therefore, appropriate for us to seek ethical and even legal recourse where possible, in matters of ecological crimes. It follows that to commit a crime against the natural world is a sin. For human beings to cause species to become extinct and to destroy the biological diversity of God's creation; for human beings to degrade the integrity of the earth by causing changes in its climate, by

Practical Sustainability Strategies: How to Gain a Competitive Advantage,
First Edition. Nikos Avlonas and George P. Nassos.
© 2014 John Wiley & Sons, Inc. Published 2014 by John Wiley & Sons, Inc.

stripping the earth of its natural forests, or by destroying the wetlands; for human beings to injure other human beings with disease; for human beings to contaminate the earth's waters, its land, its air, and its life with poisonous substances—all of these are sins" [1]. The Holy Bible goes further in stating "If we destroy the earth, then God will destroy us" [2].

As we deplete our natural resources and the population continues to increase, what is in store for human life on this planet? China's economy is growing as is their population, while the availability of water is decreasing. This means that this country and other ones may have to import more food. But how are we going to feed this growing population? A new business paradigm is necessary.

In the late 1980s and the early 1990s, a business concept was introduced called Total Quality Management (TQM). This was promoted excessively in journals and by consultants. This eventually led to other business strategies like lean manufacturing, Six Sigma, and just-in-time inventory. These strategies were imbedded in such a way that they became standard operating procedure (SOP).

Today, more than ever, it is critical to adopt sustainable strategies as described in earlier chapters as well as practices that are considered "green" but are not necessarily sustainable. Included in this category of what can be called "less bad" are energy efficiency, pollution prevention, waste reduction and recycling, supply chain management, and others. All of these can still be considered critical.

But then again, how does one know whether they are going in the right direction? This is where measuring certain key performance indicators (KPI) is important. And being able to communicate this information to the business community, the NGOs, the government, and, most of all, to the consumer is very important.

Sustainable strategies are needed more and more in organizations today, and the old business model with concentration on shareholder value seems to be replaced slowly by the stakeholder value concept. We need to move on from the short-term strategies to long-term ones, the core element of which will be more balanced relationships between communities, the environment, and economic performance. Perhaps in 10 years we will not be talking about sustainability since all the organizations will have already implemented it just as it happened with the TQM concept.

Adopting sustainable strategies and measuring and reporting the results has become and will continue to be an important dimension of the business world and, more important, the global environment for the next few decades. While it will become the new norm, the importance of this business philosophy will only diminish after we are assured that the "environmental cliff" is not in the horizon.

REFERENCES

1. Bartholomew, Ecumenical Patriarch of Constantinople, Chryssavgis J. Environmental symposium, November 8, 1997. In: Fr. Chryssavgis J, editor. *Cosmic Grace and Humble Prayer: The Ecological Vision of the Green Patriarch Bartholomew*. Grand Rapids, MI: W.B. Eerdmans Publishing Company; 2003. p. 220.

2. Holy Bible, *Revelations* 11:18.

Additional Case Studies for a Practical Understanding of Sustainability

IKEA

Related to This Case Study [1]: Chapter 4

Overview While this case study relates to the company's earlier years, it presents an excellent description of how this firm committed to The Natural Step (TNS) in order to grow into an excellent and sustainable company.

IKEA, a Swedish home furnishing retailer, is known as the world's largest designer and retailer of well-designed, inexpensive, and functional furniture for homes. The company is owned by a nonprofit foundation and has grown 15% per year in this decade (Fiscal Year 1990–1997 average growth rate). Each year, IKEA has over 140 million visitors to the 140 stores in 29 countries and distributes over 80 million IKEA catalogs. IKEA designs all 11,000 items in the product line. Product manufacturing occurs both at IKEA production facilities and at approximately 2400 suppliers in 65 countries. Today, employees number 36,400 and sales for Fiscal Year 1997 were $5.86 billion (US dollars).

In 1990, IKEA adopted TNS Framework as the basic structure for implementation of its environmental policy and plan. Using TNS principles and system conditions, IKEA has made a number of changes affecting its products and services. This case describes many of the results of these changes, along with the issues and events that lead IKEA to adopting the TNS Framework and formulating an environmental plan.

Practical Sustainability Strategies: How to Gain a Competitive Advantage,
First Edition. Nikos Avlonas and George P. Nassos.
© 2014 John Wiley & Sons, Inc. Published 2014 by John Wiley & Sons, Inc.

Background

IKEA was founded in 1943 by 17-year-old Ingvar Kamprad. As a young entrepreneur in south Sweden, Kamprad soon turned his business into a mail order operation selling a variety of household products, particularly furniture. The first IKEA showroom/store opened in 1953 in Sweden.

Kamprad's innovative strategy was to design functional furniture that was easy and inexpensive to build, receive it disassembled at stores, and display it on the showroom floor with detailed explanation tickets, making salesperson assistance unnecessary. Employees were available for questions, but the customers could choose, order, pick up, transport, and assemble their own selections. Cost savings earned by IKEA were passed through to customers in lower prices (estimated cost savings are 20–50%, compared with the competition). His stores soon became home furnishing centers, also offering restaurant facilities and play areas for children. The strategy continues to drive IKEA operations.

From the start, Kamprad's desire to integrate social value into business practice has strongly influenced the IKEA vision. In December 1976, Kamprad wrote, "What is good for our customers is also good for us in the long run." This objective of responsibility drives the company vision to create a better everyday life for the majority of people. The vision is realized by offering a wide range of functional and well-designed home furnishing items, at prices so low that the majority of people can afford to buy them.

An Environmental Challenge

In the mid-1980s, IKEA ran into an environmental problem that had significant implications on the firm's furniture line. Tests on some IKEA particleboard furniture products showed that formaldehyde emissions exceeded the standard specified by Danish environmental law [2].

Obviously this situation created a huge problem for IKEA, given the extensive use of particle board in IKEA furniture products. If the particle board from one product violated the standard and was deemed hazardous, then all products using particle board could be deemed hazardous. Negative publicity required a quick response.

While IKEA was searching for solutions, new German environmental law was announced that required formaldehyde emissions from particle board to not exceed 0.01 parts per million (the German E1 standard). IKEA elected to apply the E1 standard, the strictest in the world, to all markets by requiring that all of its particle-board suppliers meet that standard.

This long-term solution proved beneficial when California voters passed Proposition 65, tightening formaldehyde emissions and prosecuting stores selling products exceeding the standard. IKEA avoided the costs of litigation and retooling or revamping the product line because its company-wide formaldehyde requirement exceeded the California requirement. A visit by company executives to the California Attorney General, to inform him of the IKEA standard, even eliminated the cost of investigation.

In the late 1980s, IKEA and other European retailers were receiving pressure, including calls for boycotts from environmental groups, to eliminate the use of tropical rainforest wood in furniture. These pressures made it clear to IKEA-Group Executives that environmental issues would impact the future credibility of IKEA. Therefore, CEO Anders Moberg, who was personally concerned about the pace and extent of environmental deterioration, appointed Russel Johnson as the manager responsible for environmental issues.

Commitment to TNS Framework

In 1990, Johnson invited Karl-Henrik Robèrt, founder of TNS, to speak at an internal ECO seminar with the Board of Directors. Dr. Robèrt was viewed as having a new approach to environmental issues. Whereas other environmental groups were good at describing environmental problems, TNS offered clear guidance on how the problems involved IKEA and what the company could do about them from both a strategic and operational point of view. Based on the awareness created by the TNS four system conditions, the relationship with TNS developed into a commitment to work with Dr. Robèrt to develop an environmentally friendly business and contribute to a sustainable society.

Throughout the spring of 1990, a series of group management meetings produced an environmental policy that the IKEA board approved in August 1991 (Section III in tool kit). The implementation and training program of the policy are based on TNS system conditions.

In many ways, the historical values of the company were a natural basis from which to accept the TNS system conditions and adapt business operations. For example, Kamprad had viewed the minimal use of resources essential to keeping a "low-price picture." Furthermore, he valued innovation in employees and encouraged responsibility and decision-making at all levels of the organization. Therefore, as the company began to develop an environmental program, it became a natural extension to the corporate culture. In keeping with the IKEA vision, Anders Moberg, CEO, wrote, "Once and for all, IKEA has decided to side with the majority of people: to create a better everyday life. Therefore, it is our responsibility to do what we can to contribute to a better environment" [3].

The IKEA Environmental Program

In 1992, the environmental policy was transformed into an Environmental Action Plan describing concrete and practical measures for the mid-1990s. As part of the plan's development process, 25 top managers attended a two-day seminar with presentations given by Karl-Henrik Robèrt, the president for Swedish Greenpeace, an environmental legislative expert, and other environmental speakers. Following the presentations, the managers discussed a proposal for the Environmental Action Plan. Working groups were formed to agree upon the detailed activities for the plan.

The plan is a living document and is periodically updated. Unit managers receive the plan and decide how to focus implementation efforts in their business units. Specific implementation tasks fall into six categories: management and personnel; products and materials; customers; suppliers; buildings, equipment, and consumable materials; and transport.

IKEA seeks to achieve substantial environmental improvements by focusing implementation efforts on structural changes, those that impact processes, methods, or material content. By keeping the efforts focused on structural change, IKEA can maximize the impact of resources invested and reduce the energy necessary to address isolated issues. Some examples of structural changes include (i) the use of the E1 standard for all IKEA products in all sales markets, (ii) the use of ultraviolet (UV)-hardened and water-based lacquers to avoid solvents, and (iii) the process of optimizing transports to reduce exhaust emissions. In a number of cases, the efforts have resulted in long-term cost reductions.

The following sections highlight many results from the six implementation areas.

Management and Personal

This category recognizes the crucial need for individual contribution to successfully realize the environmental policy. Key tasks involve training and communication. Manager training addresses specific issues or problems in the manager's functional area. Coworker training includes general information about environmental issues and the IKEA environmental policy and action plan and "function-specific" information about the known environmental problems related to the functional area of the employee.

Results Achieved

1. By 1995, IKEA North America implemented an environmental training program, with TNS principles at its core. The training program utilizes the "train-the-trainer" principle. In the first step, the trainers are selected

from different organizations and functions within the company and then trained at a five-day seminar. In the next step, these trainers are assigned to educate all management teams and all employees, primarily those having a direct customer or supplier contact. For each group, the extent of the program is adapted to the functional needs. The basic modules include:

(a) Basic environmental knowledge according to TNS

(b) The company's environmental program: background, policy, and action plan

(c) Education adapted to the tasks of the group (e.g., range, purchase, distribution, retail)

2. Stores receive details of the IKEA position on different environmental issues to use for addressing questions or concerns raised by customers.

3. An "ECO-facts" database was created that contains brief descriptions of different topical environmental issues with summaries of known facts (see Exhibit A for an example entry). Coworkers have access to the "ECO-facts" database to gather information to address customer inquiries or solve other problems.

4. Some coworkers have voluntarily started local environmental working groups.

Products and Materials

This category recognizes that products and packaging must convey a clear signal about the commitment to the environment. Key tasks involve evaluating materials and manufacturing methods to identify the environmental impact of the materials or methods. When assessing the environmental impact of product materials, IKEA applies the environmental laws and standards from the strictest market as a minimum requirement for the products sold in all markets.

Facts about Formaldehyde

What is it? Under normal conditions, formaldehyde is a colorless gas with a pungent smell. It occurs naturally in all living cells and therefore also in the human body. Formaldehyde is able to combine with a number of substances to form a variety of end products, and synthetically manufactured formaldehyde is used in the manufacture of paints, lacquers, adhesives, rigid plastics, and a number of toiletry items, such as shampoo and soap. Formaldehyde is normally used in bound form or in aqueous solution as formalin. Formaldehyde also occurs as a by-product of incomplete combustion, for example, in car exhaust fumes and tobacco smoke where it is present in much higher concentrations than emitted from, for example, furniture.

How is the environment affected? Formaldehyde is quickly broken down in nature and is not accumulated in animals and plants. Formaldehyde can, however, cause allergic reactions when in contact with skin or if inhaled. In very high doses over a long period of exposure, formaldehyde is suspected of being carcinogenic. There is, however, no scientific evidence for this.

Is it used in IKEA products? Formaldehyde occurs in IKEA products as a binder in wood-based materials such as particleboard, bentwood, and plywood. It also occurs in adhesives and lacquers, and in textile materials as a component in finishing treatments.

What rules apply generally? IKEA has long worked to minimize the use of formaldehyde. Since 1986, IKEA has been applying the German formaldehyde requirement, currently the strictest, for all IKEA products on all sales markets, even where no limit exists. The German limit is such that even persons who are oversensitive to formaldehyde should not experience any problems. Denmark and Austria have similar requirements, while Sweden, Norway, Finland, and California have their own formaldehyde requirements.

On its own initiative, IKEA has also introduced equivalent requirements on textiles in spite of the fact that formal requirements exist only in Japan and Finland.

Results Achieved

1. Polyvinyl chloride (PVC) is gradually being phased out at IKEA. It has been replaced in wallpapers, home textiles, shower curtains, lampshades, and furniture. PVC has also been eliminated from all packaging and is gradually being phased out in electric cables.

2. IKEA is at the forefront of minimizing the use of formaldehyde in its products, including textiles (despite the fact that formal requirements for formaldehyde in textiles exist only in Japan and Finland).

3. Acid-curing lacquers have been replaced with alternatives (e.g., UV-hardened and water-based lacquers).

4. A version of the IKEA OGLA chair is made from 100% recycled preconsumer plastic waste.

5. A product called "a.i.r.," consisting of a series of air inflatable furniture products (e.g., a sofa), has recently been introduced into the product line. Individual components are inflated by the customer, using a hairdryer, and then individually "stuffed" into a slipcover that serves as the item's frame. The resulting product reduces the use of raw materials for

framing and stuffing. In addition, transportation weight and volume are reduced to about 15% of what is required for a conventional sofa.

6. In powder lacquers, the use of chromium for metal surface treatment is being reduced.

7. The use of substances such as cadmium, lead, PCB, PCP, and AZO pigments is prohibited or strictly limited.

8. IKEA strives to use only wood from known, well-managed sources: forests that replant and care for the protection of biological diversity.

9. IKEA uses only recyclable materials for flat packaging. In addition, using "pure" (nonmixed) materials for packaging enables easy sorting/recycling.

Customers

This category recognizes the need to make it easy for customers to incorporate environmental considerations into purchase decisions. Tasks seek to give customers sound environmental information and provide environmentally friendly alternatives for acquiring IKEA products.

Results Achieved

1. In 1992, IKEA worked with Greenpeace to develop guidelines for catalog production. Today, over 80 million IKEA catalogs are printed on non-chlorine-bleached paper and use pulp from farmed wood (no old growth). In addition, the company issues only one catalog per year, utilizes 10–20% postconsumer recycled paper, and accepts old catalogs back at stores for recycling. Additional environmental highlights of catalog production include the use of:

    • Digital engraving at the print shop, rather than traditional film reproduction. This process reduces the use of plastic film and heavy metals, and there are no chemicals in the reproduction process.

    • Low-toluene content ink and heavy-metal-free ink, resulting in less use of solvents.

    • Adhesives that are free from injurious chemicals.

2. Several European stores offer free bus transportation from selected city areas to the store, as an alternative to use of private cars. When public bus transportation became available to a German IKEA store, 33,000 additional individuals visited the store in the following year.

3. Stores will accept product packaging that the customer wants to leave.

4. Organic cotton fabrics are available for custom upholstery.

Suppliers

This category recognizes the need to encourage suppliers to adopt environmentally responsible production methods. Key tasks are to encourage suppliers to strengthen their awareness of environmental issues and to support the development of more environmentally sound production technologies.

Results Achieved

1. The process of working with suppliers for the North American market has been challenging. Robert Paolozza, IKEA NA manager responsible for quality and environmental issues, said that most suppliers were "somewhat surprised" at the environmental requirements of the IKEA product specifications. It has been necessary to work closely with suppliers to help them understand and adapt to key environmental product specifications, including restrictions on formaldehyde, lacquers, and wood sources (no rainforest wood). Also, packaging has to be recyclable or reusable and contain no PVC.

2. In Northern Europe, IKEA has organized 2½-day environmental workshops for suppliers. The workshop is offered to suppliers at cost. Participating companies send one or two individuals to be trained to train others on environmental issues. Later, the participators are prepared to conduct training at their own company and to help establish their own environmental program.

3. Several IKEA suppliers in European countries now act according to established environmental standards and use an environmental management program. Standards used for certification are ISO 14001 or the Eco-Management and Audit Scheme (EMAS), a European Union regulation. Many more suppliers are in the certification process.

Buildings, Equipment, and Consumable Materials

This category recognizes the need to work for a better environment in all "we" do. Efforts in this area include evaluating the environmental impact of property, property enhancements, waste, equipment, and materials. For example, the environmental impact of office machines and materials is evaluated, and if more environmental compatible alternative products are available, they should be chosen at the time of next procurement. Similar practices are also used for other kinds of equipment (e.g., forklifts) and consumable supplies.

Results Achieved

1. Newly built IKEA stores and other owned property are constructed according to environmentally adapted requirements. Every effort is also made to adapt to these requirements when renovating old property.

2. Many European IKEA stores have adopted a "Trash is Cash" program. Trash is Cash takes IKEA packaging materials (e.g., cardboard, plastics), recyclable office waste (e.g., paper), and other store waste (e.g., paints, glass, wood) and recycles it.

3. In 1993, the Gothenburg, Sweden, store set up its own on-site recycling facility, and today the store recycles almost 85% of its waste. Its annual solid waste bill, about $35,000 (US dollars) in 1992, is now a small profit. On-site recycling facilities are now established at all Swedish IKEA sites (including stores, offices, and warehouse/distribution centers). Waste is sorted into 16–22 fractions, and 80–85% of total waste volume is sorted. The program considerably reduced waste handling costs; total cost today is close to zero. The goal is to reach 100% sorting/recycling within a few years. There also is a prototype in use for on-site composting of restaurant waste.

4. In Switzerland, stores offer customers the ability to deposit old furniture when replacing it with new IKEA furniture. By depositing old furniture with IKEA, customers can save about half of the waste disposal cost (e.g., nearly $100 for a sofa). A recycling contractor dismantles the furniture and sorts the materials into different fractions: wood, metal, textile, plastic, etc. The IKEA goal is to offer this service at a break-even cost point.

5. In 1995, the Philadelphia store was retrofit with fluorescent lighting at a one-time cost of $151,000 (US dollars) and expected yearly savings of $85,322 (US dollars) through less energy (in KWH) use. By the end of 1998, all IKEA North America facilities will be retrofit with fluorescent lighting.

6. Some buildings are experimenting with alternative energy sources (e.g., photovoltaic solar systems and use of groundwater to heat/cool the indoor climate).

Transport

This category recognizes the need for environmentally sound transportation methods. Efforts seek to reduce the demand of nonrenewable natural resources such as oil and direct damage to the environment as a result of emissions. Specific tasks achieving positive results over time include:

- Using flat IKEA packaging that takes up little transport volume.
- Using railroads for long-distance transportation.

- Maximizing the efficiency of shipments: reduce the number of transports and the number of empty transports, make maximum use of cargo vehicle space, utilize return transportation, and avoid rush-hour traffic.
- Choosing transportation companies that meet EC standards on emissions and noise.

Results Achieved

IKEA has continually applied logistic solutions to all distributed goods (e.g., product, catalog, and fixtures). These items have resulted in real cost savings for IKEA, through the reduction of waste and the efficiency of transport. For example:

- Recycling shrink film.
- Using returnable pallets.
- Using combi-transports, that is, goods are conveyed by rail for part of their journey and by road for the remainder of the journey.
- Using transportation units in both directions when possible.
- Creating a "smart" goods logistics, for example, using a bookbinding contractor between the printing house and the distribution center, to minimize total transportation distance.

Benefits, Challenges, and Lessons Learned

To date, IKEA has not focused on measuring tangible benefits of its environmental program. Plan implementation costs have been viewed as operational or product costs. Intangible benefits have affected the employees, customers, suppliers, and product line.

The environmental training program received a fantastic response from employees and good support from management. Employees are motivated to work for a company with an environmental commitment.

Consumers, in all markets, are benefiting from the IKEA adherence to strict environmental standards, regardless of the regulations in that market.

Supplier relationships are strong. IKEA has worked closely with suppliers to help them modify production processes to meet revised product specifications. Such modifications have often resulted in production efficiencies and a reduction in total product cost.

IKEA has made the strategic decision to focus its energy internally on continuous improvements that support the environmental policy and plan. Therefore, external communication of the plan's implementation is made through "proving results."

The greatest challenges that lie ahead are in the areas of sustainable forestry, producers' responsibility legislation (Sweden and Germany), and transportation. The following insights share some lessons learned from the IKEA experience:

- Create awareness by involving as many people as possible from the start.
- Align your environmental work with your business vision; it must fit your business reality.
- Keep it simple in simple words!
- Put the environmental issue deep into the line organization—don't departmentalize it; it concerns everybody.
- Start with visible actions that show concrete results.
- Have a champion, someone whose job is to focus on the key issues. Managers and coworkers will absorb the "functional view"; a champion can advocate a "systemic view."

"Even on a day-to-day operation," says Russel Johnson, director of environmental affairs, "there is a lot to be done." Basic tasks that will help achieve objectives include the following:

- Avoid complicated specialist terms by using "every-man" wording and explanations.
- Try to find "down-to-earth" changes and solutions. Communicate to employees and others involved.
- Encourage employee volunteer activities and behavior changes at home as well as at work.

Issues for Learning and Discussion

1. What were the IKEA sustainability strategy and its key impact areas?
2. What was the impact of adopting TNS?
3. What were the environmental and social benefits?
4. What were the challenges and benefits of its strategy?
5. What lessons were learned for the future?

LLOYDS BANKING GROUP: TRANSLATING SUSTAINABILITY VISION INTO STRATEGY

Related to This Case Study [4]: Chapters 15–17

 *Lloyds Banking Group (LBG)* is a leading UK-based financial services group providing a wide range of banking and financial services to personal and corporate customers. Formed through the acquisition of HBOS by Lloyds TSB in 2009, it is now the 23rd-largest company on the London Stock Exchange, with more than 110,000 employees serving the financial needs of over 30 million customers. Financial services are offered through a number of well-recognized brands including Lloyds TSB, Halifax, Bank of Scotland, and Scottish Widows.

The *vision* for LBG is to be recognized as the best bank for customers. Its *strategy* is to create value for shareholders by investing where it can make a real difference for customers, communities, and colleagues.

Trying to preserve a vivid *vision* for the future, LBG in collaboration with the Centre for Sustainability and Excellence (CSE) embarked on a sustainability journey and since 2004 has identified and responded to great challenges. Centre for Sustainability and Excellence has provided a variety of consulting services to LBG according to CSE methodology.

Stakeholder Mapping and Engagement

The identification of all stakeholders and the understanding of their needs and expectations is a crucial starting point for organizations that want to be socially responsible. LBG knows that it can only achieve its aim of being the best bank for its customers by demonstrating that it is acting in their best interests. This means proactive engagement, listening, and making changes to the way they do business.

LBG CSR strategy is to support the corporate vision by helping to build a great place for people to work, a great place for customers to do business, and generating great returns for shareholders. LBG creates value for all stakeholders on an ongoing basis, through:

- Increased employee engagement and customer satisfaction
- Enhanced brand perception, consideration, and commitment
- More effective risk management
- Improved responsiveness to changes in patterns of customer behavior

- Supporting development of new markets and innovation in existing markets
- Delivering competitive advantage through better corporate responsibility management

CSR Training and Awareness

Investing in people is crucial to staff engagement and business growth. In order to deliver corporate responsibility performance that reflects the key strategic priorities of the business, LBG is undertaking trainings depending on the needs of employees. More specifically:

- Annual Sustainability Assessment based on the European Foundation for Quality Management (EFQM) framework for CSR.
- Easy access to a range of support services and training materials through intranet facilities.
- Forty thousand customer-facing employees receive additional training on handling complaints LBG.
- Two thousand and five hundred staffs trained to open business accounts and advice start-ups.
- Regular employee training for understanding financial crime risks (zero-tolerance approach to bribery and internal fraud).
- Regular employee training in environmental risk management.
- LBG launched a Banking Group health and safety training suite in order to achieve "world-class" standards in managing health and safety.

In 2010, LBG provided an average of 5.4 days of formal learning per FTE, a significant increase from 2.9 days in 2009. In the group-wide Colleague Engagement Survey, 77% of employees said they believed they had the opportunity for personal development, and 78% said there are sufficient opportunities for them to receive training to improve their skills. In 2011, LBG focus is on embedding the new model, training, and standards.

Key to the success of LBG are written commitments on creating a great place to work, which entails employee engagement strategy and executive development, leadership, and management programs, for the development of leadership and capability. What is more, diversity and inclusion is central to the business success for LBG, in terms of gender equality, ethnic diversity, disability, sexual orientation, and generational diversity.

Annual CSR Review and Assessment

LBG uses the EFQM's "Corporate Responsibility Framework" in order to shape their responsible business strategy and check their performance. As part of the process, LBG carries out an annual self-assessment of their performance in order to identify strengths and areas for improvement and set objectives and actions for the future. It is important for LBG to judge their performance against other companies, and this framework provides a benchmark to help them achieve this.

Independent consultants review LBG performance each year and give an independent assurance on the report. LBG also measures their performance against their peers by comparing it to several external benchmarks and indices: the Business in the Community CR Index, the FTSE4Good Index, and the Dow Jones Sustainability Index. Moreover, LBG strengthened its governance framework in 2010, establishing board representatives for key strands of their responsible business agenda. The board considers responsible business issues throughout the year and reviews their performance on an ongoing basis.

Internal Communication

Effective internal communications have played an important role in the group's performance, service delivery, and reputation. They drive employee awareness, advocacy, and engagement and enable employees to make a contribution in delivering the corporate agenda. Moreover, LBG open and transparent internal communications are rated as 20 points above the national norm. Characteristically:

- LBG ran the "Spotlight on Smart & Responsible" campaign to provide employees with information and engage them in the Environmental Action Plan.
- The Group's Sustainability Network was relaunched in a more action-focused way.
- LBG delivered monthly health and safety internal communications to all employees, highlighting risks and risk mitigation strategies.
- The group consults extensively on all proposed changes with its four recognized unions.

Integration of CSR into Policy, Strategy, and Reporting: Using the CSR Performance Scorecard

For LBG, corporate responsibility is an investment in the strategic assets and capabilities that drive business performance. LBG adopted an upgraded strategic tool, in order to measure its CSR performance, set clear goals for sustainability, implement CSR initiatives, apply new measurements, and meet goals.

By adopting a tailor-made CSR performance scorecard, LBG achieved:

- Corporate Responsibility Strategy with strategic focus on business through integration of individual business balanced scorecards
- Delivery of local programs and stakeholder engagement strategies that maximize employee involvement and engagement
- Performance and perception of performance CSR indicator analysis
- Strengths and key areas for improvement identification
- Setting of key performance indicators (KPIs) based on these priorities aligned with specific target measures under a balanced scorecard for business units and individuals
- Identification of improvement priorities and improvement activity that align with the business balanced scorecard objectives for each business unit
- Frequent self-assessments involving key enablers from across the business against the CSR criteria
- Sharing results and progress with employees throughout the business, to highlight achievements and gain commitment and engagement in improvement activity

This approach has allowed LBG to:

- Identify risks and opportunities through the CSR assessment
- Implement targeted CSR actions to achieve the corporate and sustainability goals
- Communicate the CSR actions and the company's performance to all its stakeholders
- Address mutual economic, environmental, and social challenges by improving dialog with its stakeholders

Adding Credibility to the Annual Sustainability Report: CSR Report Development and External Quality Assurance

In order to identify areas for improvement that have been built into the balanced scorecard priorities, a number of representatives from key business units contributed to the corporate responsibility report. LBG adopted a sustainability framework based on the key principles of sustainability, in order to develop a report to meet needs and expectations of a wide range of stakeholders. In addition, an external assurance statement provided an independent verification and a comprehensive validation of the report content and therefore added value to the quality of the report.

Additionally, feedback was provided for further development of the CR Report content, based on the CSE good practice research. As a result:

- CSR actions and company's performance were communicated to all stakeholders.
- Mutual economic, environmental, and social challenges were addressed by improving dialog with its stakeholders.

Advanced CSR Report Based on the Global Reporting Initiative (GRI) Framework

In 2010, LBG used for the first time the internationally recognized GRI G3 Sustainability Reporting Guidelines and the financial sector supplements in the development of the CSR report.

The 2010 Responsible Business Report presents the commitments and the achievements of LBG, in relation to its overall economic, environmental, and social performance as well as all related improvements made within this period for the entire UK operations. LBG has self-declared the 2010 Responsible Business Report to be Application Level B+. Centre for Sustainability and Excellence has checked and confirmed that the 2010 LBG has satisfactorily applied the GRI Reporting Framework, as specified by G3 Guidelines. The 2010 Responsible Business Report meets the requirements of GRI G3 Version 3.0 Application Level B+.

Measurements and Impacts LBG measures its performance, identifies challenges and opportunities, and reports under three main headings: responsible business management, economic and social impact, and building relationships.

1. *Responsible Business Management* LBG approach to responsible business management is founded on robust corporate governance practices and a risk management culture, which guides the way all employees approach their work, the way they behave, and the decisions they make.

Results in 2010

- £60.6 billion reduction in LBG reliance on liquidity support from government and central bank.

- Ninety percent of employees work on harmonized employment terms and conditions.
- 300,000 customers/month contacted to assess their financial health.
- £12.5 million were donated to money advice and debt charities.
- Smart and Responsible initiative launched to reduce LBG environmental impacts.
- £5 million invested in energy-saving and energy efficiency schemes.

LBG's new strategy aims to deliver its vision of making a real difference to the United Kingdom. What is more, LBG will strengthen the governance framework by refreshing the group's code of business conduct.

2. *Economic and Social Impact*   While its main contribution to society is its direct economic impact, LBG strongly believes that this must also be supported by its active investment in the communities in which it operates. Its performance regarding the economic and social impact shows LBG's strong sense of obligation towards society and economic recovery:

 - 100,000 new business start-ups financed
 - £79 billion of new lending to homeowners and businesses
 - £30 billion of gross new mortgage lending to UK homeowners
 - £11 billion of new lending to small- and medium-sized enterprises
 - £5 billion of new lending to first-time home buyers
 - Thirty-two percent share of all new social bank accounts opened during the year
 - £76 million invested in UK communities
 - £480 million given to the group's charitable foundations over the past 25 years

3. *Building Relationships*   Building strong relationships with stakeholders and helping them to develop is fundamental to the success of the business and for achieving the vision of being the best bank for their customers. Measures as well as impacts used to monitor and evaluate performance include the following:

 - Seventy-five thousand customers contacted every month as part of customer feedback process.
 - Ninety percent of customer complaints are resolved straightaway.
 - Free mobile banking for all LBG TSB customers.
 - Money Manager online tool launched to help customers track spending patterns.

- 900,000 courses on the new learning at LBG website.
- Six hundred business and environment managers trained.
- Ninety-three percent of employees say they work beyond what is required to help LBG succeed.

First bank to communicate with customers on Twitter and YouTube

Sustainability Achievements   LBG has the following achievements:

- The only bank listed as a Platinum company in Business in the Community's Corporate Responsibility Index.
- Ranked as the top UK bank in the FTSE4Good Index.
- The top UK bank in the new Carbon Disclosure Project (CDP) Carbon Strategy Index Series.
- One of a select group of companies to hold Business in the Community's Community Mark—the national standard that publicly recognizes excellence in community investment.
- Included in the Dow Jones Sustainability Index and in the Carbon Disclosure Leadership Index.

Issues for Learning and Discussion

1. What are the main elements of the stakeholder engagement strategy for LBG?
2. What are the key success factors for the implementation of CSR in LBG?

APPLE [5A, B]: SUPPLY CHAIN RISKS

Related to This Case Study: Chapters 14, 15, and 20

Responding to a growing outcry over conditions at its overseas factories, Apple announced that an outside organization had begun to audit working conditions at the plants where the bulk of iPhones, iPads, and other Apple products are built and that the group would make its finding public.

For years, Apple has resisted calls for independent scrutiny of the suppliers that make its electronics. But for the first time it has begun publicly divulging information that it once considered secret, after criticism that included coordinated protests last week at Apple stores around the world and investigative news reports about punishing conditions inside some factories.

In January 2012, Apple released the names of 156 of its suppliers. Two weeks later, Apple's chief executive sent an e-mail to the company's 65,000 employees defending Apple's manufacturing record while also pledging to go "deeper into the supply chain." And now, the company has asked an outside group—a nonprofit financed partly by participating companies like Apple—to publicly identify specific factories where abuses are discovered.

Corporate analysts say Apple's shifts could incite widespread changes throughout the electronics industry, since a lot of companies use the same suppliers. They also said it seemed calculated to forestall the kind of public relations problems over labor issues that in previous decades afflicted companies like Nike, Gap, and Disney. "This is a really big deal," said Sasha Lezhnev at the Enough Project, a group focused on corporate accountability. "The whole industry has to follow whatever Apple does."

But it is unclear if the efforts by Apple, whose $469 billion market value is the largest of any company in the world, will be enough to quiet its critics, some of whom had urged Apple to work with Chinese monitoring organizations with direct knowledge of its suppliers in China.

Though some labor groups applauded Monday's announcement, others said that the outside auditor Apple chose, the Fair Labor Association, which is based in Washington, was not sufficiently independent. And some critics questioned whether the inspections—Apple said the manufacturers had agreed to do them voluntarily—would sharply curtail problems or merely help Apple deflect criticism.

> "F.L.A. is part of a corporate social responsibility industry that's totally compromised," said Judy Gearhart, executive director of the International Labor Rights Forum, an advocacy group for workers. "The auditing has been proven to be weak, and real solutions need a lot more than auditing. It takes empowering workers"(http://bits.blogs.nytimes.com/2012/02/13/apple-announces-independent-factory-inspections/).

Apple, in a statement, said that the Fair Labor Association was an independent organization that had been given "unrestricted access" to the company's suppliers. The first inspections, Apple said, were conducted on Monday at a factory in Shenzhen, China, known as Foxconn City, one of the largest plants within China. Human rights advocates have long said that Foxconn City's 230,000 employees are subjected to long hours, coerced overtime, and harsh working conditions, all of which Foxconn disputes.

Apple has said that if the companies manufacturing its products do not measure up to its labor and human rights standards, it will stop working with them.

"We have a very credible, independent monitoring system," said Jorge Perez-Lopez, executive director of the Fair Labor Association. "Yes, Nike is on our board. So are other companies. But so are universities. And our reports are written by staff, without consultation or influence" (http://www.apple.com/supplierresponsibility/labor-and-human-rights.html).

Since 2007, Apple has released yearly audit reports of its own, detailing labor violations and unsafe conditions at its suppliers. More than half of the facilities audited by Apple every year had violated at least one aspect of the company's supplier code of conduct and in some instances violated the law.

Auditors have found instances of excessive overtime, underage workers, improperly disposed hazardous waste, and falsified records, according to the company's reports. Two years ago, 137 workers at an Apple supplier in eastern China were injured after they were ordered to use a poisonous chemical to clean iPhone screens. Last year, two explosions at iPad factories killed 4 people and injured 77.

Because Apple's public disclosures of problems do not identify the suppliers by name, it is difficult to determine where specific abuses happened. It is also tough to determine if conditions improved after Apple demanded changes, as the company says. The company has disclosed little about its manufacturing process over all, at least partly because it does not want to tip off competitors.

For instance, Chinese advocacy groups—which are often considered reliable, independent monitors—have published multiple reports saying that Foxconn employees regularly work more than 12 hours a day, 7 days a week, a violation of both Chinese law and Apple's code of conduct. Apple has audited Foxconn City multiple times, and Foxconn, in a statement sent to *The Times*, said it had never been cited by Apple for overworked employees.

If the Fair Labor Association conducts wide-ranging audits and publishes data on specific facilities, it could transform the attention brought to the worst performers and, in the example of Foxconn City, help determine whether Foxconn or the advocacy groups have been telling the truth.

"The problem with the F.L.A. is that it lives by rules set up by the companies itself," said Mr. Lezhnev of the Enough Project. "Real transparency will transform the electronics industry. But if it's just a whitewash, I'm not sure how much will change" (http://bits.blogs.nytimes.com/2012/02/13/apple-announces-independent-factory-inspections/).

Apple, in its statement, said the Fair Labor Association's findings and recommendations from its first inspections would most likely be posted online in early March on the group's website, fairlabor.org.

At Apple's request, the group will also conduct audits of Apple's other main assembly factories, including Foxconn's plant in Chengdu and facilities run by Quanta and Pegatron, where the bulk of iPhones, iPads, and other devices are made. Those and other plants also build goods for almost every other major electronics company, including Dell, Hewlett-Packard, IBM, Lenovo, Motorola, Nokia, Sony, and Toshiba.

While other companies have been criticized for conditions at their operations overseas, Apple has received particular attention because it is the biggest—its market value is more than the combined value of Google and Microsoft—and among the richest. Its stock closed Monday at $502.60, up more than 20% this year. The company also has a vast overseas presence, with its contractors employing 700,000 people in China and elsewhere.

Apple announced in February 2012 that its suppliers had pledged to give the auditors unrestricted access to their operations during the inspections. Apple said the organization would "interview thousands of employees about working and living conditions including health and safety, compensation, working hours and communication with management." It will also inspect manufacturing areas, worker dormitories, and other facilities, the company said.

> "We believe that workers everywhere have the right to a safe and fair work environment, which is why we've asked the F.L.A. to independently assess the performance of our largest suppliers," Timothy D. Cook, Apple's chief executive, said in a statement. "The inspections now under way are unprecedented in the electronics industry, both in scale and scope, and we appreciate the F.L.A. agreeing to take the unusual step of identifying the factories in their reports" (http://bits.blogs.nytimes.com/2012/02/13/apple-announces-independent-factory-inspections/).

In January 2012, Apple announced it would be the first technology company to join the Fair Labor Association. The organization was founded in 1999 and evolved out a task force created by President Bill Clinton and a handful of apparel and footwear companies—including Nike—to combat child labor and other abusive working conditions.

When completed, Apple said, the association's inspections will have covered factories where more than 90% of Apple's products are assembled [1].

> In announcing that the association had begun inspecting Foxconn factories in Shenzhen and Chengdu, Timothy D. Cook, Apple's chief executive, said, "We believe that workers everywhere have the right to a safe and fair work environment, which is why we've asked the F.L.A. to independently assess the performance of our largest suppliers" (http://www.apple.com/pr/library/2012/02/13Fair-Labor-Association-Begins-Inspections-of-Foxconn.html).

Critics argue, however, that the association and its corporate members should not suggest that its inspections are independent.

> "The F.L.A. does some good work, but we don't think it's appropriate for them to call themselves independent investigators because they're in part funded by companies," said Scott Nova, executive director of the Worker Rights Consortium, a university-backed factory monitoring group. "Independent monitoring means you're generally independent of the companies" (4, http://www.nytimes.com/2012/02/14/technology/critics-question-record-of-fair-labor-association-apples-monitor.html?_r=0).

Issues for Learning and Discussion

1. What would the risks be if Apple did not provide this info to its stakeholders on time?
2. Is the Apple approach for an independent scrutiny with FLA correct?
3. Should companies like Apple put in place comprehensive sustainability (CSR) policies and provide transparent results to their stakeholders as a proactive behavior (e.g., stakeholder dialog) or just as a reaction to stakeholder pressures?

RIM: CARBON FOOTPRINT AND LCA REPORTING TO ITS STAKEHOLDERS

Related to This Case Study [6]: Chapter 18

RIM has been participating in the CDP since 2009. Its reporting is based on the Greenhouse Gas Protocol and includes Scope 1, 2, and 3 emissions. Copies of the report can be found at www.cdproject.net. RIM is actively engaged in initiatives to increase the energy efficiency of our products, data centers, and buildings.

Renewable and Responsible Energy Use

RIM continued to purchase renewable energy from Bullfrog Power in fiscal 2012. Bullfrog Power renewable energy is generated from wind and hydro facilities that are certified as low impact by Environment Canada under its EcoLogo program. A total of approximately 1747 megawatt hour of Bullfrog Power was injected into the regional grid (which matches approximately the amount of power used by six RIM office buildings during calendar year 2011).

RIM established a utility management program for our manufacturing facility according to the ISO 14001 environmental management system guidelines. The program aims to identify and improve utility use. Since it was established in 2007, approximately 2,907,000 kilowatt hour have been saved, which equates to 628 tons of CO_2 emissions. Energy savings have been achieved through scheduled shutdowns, lighting upgrades, heat load reductions, re-insulating walls, and installing heat recovery units.

Green IT

In fiscal 2012, RIM created a Green Information Technology program focused on reducing the environmental impacts of our information technology (IT) infrastructure. Data systems virtualization has been a key focus to reduce our corporate carbon footprint. RIM increased the server virtualization percentage at major data centers to greater than 65% (65% virtualization and 35% physical hardware). This resulted in approximately 80% less power consumption. RIM also maintained a 30:1 ratio for virtual servers running on each physical server in the majority of our data centers.

RIM increased internal data storage allocation efficiency and utilization rates by 50%, eliminating the need to purchase entire data storage arrays and the power and space they would consume. For many data storage systems, they have increased utilization efficiency from 35% to 60%, and it is still increasing. RIM has also implemented higher data storage efficiency technologies for new systems into our IT infrastructure. For data storage that is required to support virtual server infrastructure, RIM introduced new storage technologies that have an improved efficiency of 85%, allowing the purchase of smaller, more efficient storage. The new virtual infrastructure also includes backup data protection, eliminating extra storage capacity that has been traditionally required for this purpose.

Improving Product Sustainability

In fiscal 2012, RIM worked with experienced sustainability consultants to conduct in-depth, baseline assessments of our sustainability policies, programs, and product development activities. The Natural Step, an international nonprofit research and advisory organization, conducted a Sustainability Life Cycle Assessment (SLCA). The SLCA provided a strategic overview of the sustainability of its products, highlighting the ecological and social impacts of current products throughout their life cycle.

To further assess the impact of its products, RIM conducted comprehensive Life Cycle Assessment (LCA) studies on the BlackBerry® Torch™ 9810 smartphone and the BlackBerry PlayBook tablet. The assessment provided an in-depth view of each product's environmental impacts at every stage in its life cycle, from the materials used in the product to production and distribution, to its use, and to the end of its useful life.

Together, the SLCA and the LCA equip RIM with information to help focus our efforts on reducing our overall environmental impact.

Life Cycle Assessments of the BlackBerry Torch 9810 Smartphone and the BlackBerry PlayBook Tablet

Results from the LCA studies helped RIM identify what the company is doing well and opportunities to continue to lessen its environmental impacts. The smaller packaging reduces material use and results in more efficient transportation. The distributed manufacturing sites also reduce the impacts from transportation, and its efficient repair and refurbishment processes contribute to lowering lifetime impacts and extending the product's useful life.

The LCA findings are also useful for identifying areas that have the highest impact on the environment, helping us make informed decisions about where to focus the company's efforts for improvement.

MATERIALS

RIM continues to review the materials used in its products to make more sustainable choices. In November 2011, RIM updated its Restricted Substances List to include several additional substances. In addition, beryllium has been banned from all new products and accessories since December 2011. RIM also eliminated a number of phthalates from all BlackBerry products and accessories. In fiscal 2012, RIM started its transition towards fiber-based packaging with the introduction of a fiber tray in certain products.

End of Life

RIM offers a variety of options for customers to responsibly dispose of BlackBerry devices that have reached the end of their useful life. Customers in the United States can download a prepaid mailing label at www.blackberry.com/recycling and mail in their used devices and accessories to be responsibly recycled.

The BlackBerry® Trade Up Program is another sustainable alternative for those looking to upgrade their smartphone, offering customers an environmentally friendly solution for their e-waste. Customers are encouraged to mail back their used devices and, in doing so, will receive credit on the purchase of a new BlackBerry smartphone. Devices returned in good condition are refurbished, which will further extend the life of the product through resale and reuse. All devices not suitable for resale are responsibly and safely recycled. This program is currently available in the United States,

Canada, Mexico, the United Kingdom, Germany, Australia, the Philippines, Indonesia, and the Cayman Islands, with ongoing efforts to include more countries.

Issues for Learning and Discussion

1. What are the standards RIM used for assessing its carbon footprint and report to its stakeholders?
2. What were the main conclusions from the LCA study done for the BlackBerry Torch 9810 smartphone and the BlackBerry PlayBook tablet?

UNILEVER: WATER FOOTPRINT STRATEGY CHALLENGES AND APPROACH

Chapters Related to This Case Study [7]: Chapter 19

Unilever's Water Footprint   The Unilever Sustainable Living Plan commits the company to halving the water associated with the consumer use of its products by 2020. Water shortages are an increasingly common concern around the world and are exacerbated by a number of factors including climate change and population increase. To grow the business sustainably, it must reduce the total amount of water used across its value chain.

It conducted a detailed study of its water footprint, which involved an assessment of approximately 1600 of its products. This has led Unilever to the development of a metric that measures the water in the product as well as the water required for its use. It did this on a "per consumer use" basis, for example, the water needed for one hair wash with shampoo. It estimates that 50% of the water footprint occurs in the consumer use of our products. The approach is to focus on areas that suffer from water scarcity.

The work conducted on its footprint has helped Unilever to develop its strategy and set targets to manage the impact.

Unilever's Strategy   The strategic approach to water covers reducing water in agriculture, manufacturing, and the water associated with consumer use:

- Water in agriculture: Unilever set a goal to source 100% of its agricultural raw materials sustainably by 2020, according to the Unilever Sustainable Agriculture Code. One of the 11 indicators in the code relates to water. The requirements of the code are shared with all of its suppliers of agricultural raw materials who must comply with the

"Scheme Rules." Unilever is working with them to minimize the water used to grow, and pollution from growing, the key crops.

- Currently, the water metric does not include water used in agriculture since there is a lack of reliable data on agricultural water use for the wide range of crops that make up the ingredients in its products. However, it is working with others to develop the understanding of this area.
- Water in manufacturing: Unilever has set a goal that by 2020 water abstraction by its global factory network will be at or below 2008 levels. It is making steady progress through continuous improvement.
- Water in consumer use: one of the goals of the plan is to halve the water associated with the consumer use of the products by 2020 in seven water-scarce countries, which represent more than half the world's population.

Unilever's footprint analysis work showed that around 44% of its domestic water footprint in water-scarce countries is associated with its personal care products (soap, shower gels, and shampoos). A further 38% comes from the laundry process—a significant proportion from washing laundry by hand.

The future strategy for water will therefore be led by the personal care and laundry categories. The main approach will be to design innovative products and tools that help consumers reduce water when doing the laundry, showering, and washing hair, combined with behavior change programs to help shift to a new habit.

Water stress is a very local issue affecting regions and countries differently. Therefore, the social, environmental, and political aspects of managing water differ greatly between countries and even within countries.

The average amount of water people use varies across the world, but increases with industrialization. The United Nations states that each person needs 20–50 liters of water per day for drinking and other basic needs. The average North American uses 550 liters daily, while in some of the poorest countries, people live on as little as 10 liters. Over one billion people worldwide do not have access to safe drinking water. In 2010, the UN stressed the importance of this precious resource as it declared safe and clean water to be a basic human right.

Some 70% of total water consumption around the world is used for agriculture. As global populations grow, so too do the demands from farming. Furthermore, access to freshwater is increasingly problematic as demand grows and water sources become polluted. Even where adequate supplies of clean water exist, they may simply be unaffordable to people on low incomes.

Climate change will exacerbate the problem of water scarcity. Where this pressure on water supplies brings communities and countries into opposition, social and political conflicts may arise.

Future Challenges   A large part of Unilever's water footprint is associated with showering, bathing, and hair washing, just as its greenhouse gas footprint is associated with heating that water for showers. To reduce water use, it will need to develop new products and tools that help people use less water.

But it has learned that while creating new product technology is important, it is not enough. It still needs to motivate people to adopt the new water-saving behavior. Unilever's Five Levers for Change methodology is helping its marketers take a systematic approach to reducing water use when rinsing laundry by hand. It plans to apply the same approach to changing habits in the shower. In the end, it will require water pricing and water metering alongside consumer education to drive the right behaviors. At that point, it will need to be ready with products and tools that help people make their water go further.

Unilever participates in the CDP Water Disclosure. This initiative aims to improve and standardize corporate water measurement and reporting and raise awareness of water-related issues. The latest 2012 request was sent to around 315 of the world's largest companies based on their water use or exposure to water risks. It was backed by 470 institutional investors representing $50 trillion in assets.

Results from the 2011 disclosure request placed Unilever in the retail, consumer discretionary, and consumer staples sector. Unlike the CDP Investor (carbon) request, companies are not ranked or scored either on the quality of their disclosures or on their performance in water management. Individual company responses are available to read on the CDP website.

Unilever's disclosure of its water use globally put it at the top of the foods sector in the report "Murky waters? Corporate reporting on water risk: a benchmarking study of 100 companies." Published in February 2010, the report was a collaboration between Ceres (a US coalition of investors, environmental groups, and other public interest organizations), financial services company, UBS, and financial information provider, Bloomberg. Unilever also ranked in the top 10 companies overall in the study.

Product Water Footprint   Its products rely on water at all stages of their life cycle. It has conducted detailed measurement and analysis of the water footprint to inform its strategy.

The Impacts   Its use of water resources is both direct and indirect. Water is used:

- By its suppliers of agricultural raw materials for growing crops
- In its factories both as an ingredient in its products and during the manufacturing process
- By its consumers when they use Unilever's products to do their laundry, shower, or cook

In the absence of a quantified value for the water used in agriculture, the biggest impacts occur in the use of Unilever's personal care and laundry products.

Measuring Our Water Footprint   Using 2008 as its baseline year, Unilever assessed the water impact of 1600 of its products in 14 countries, representing over 70% of volume sales. It conducted this assessment using a metric that measures the water in the product as well as the water required for its use. They did this on a "per consumer use" basis, for example, the water needed for one hair wash with shampoo.

The water metric excludes the water used in the manufacturing operations, which is measured as part of its eco-efficiency in manufacturing program. It also excludes the water used in agriculture—it is making some progress on this but has more to do.

It chose to base its analysis on seven countries that were defined as water scarce. These are China, India, Indonesia, Mexico, South Africa, Turkey, and the United States, representing around half of the world's population. In the company's definition of water scarcity, it evaluates how many people in each country experience physical water scarcity as well as the number of people who have access to an improved water source. In November 2010, as part of its Sustainable Living Plan, it set an ambitious target to halve the water associated with the consumer use of its products by 2020.

Understanding Its Footprint   Unilever's analysis has helped to see which of its product categories are more water intensive than others and therefore yield the biggest opportunities for reductions. For example, around 44% of Unilever's domestic water footprint in water-scarce countries is associated with its personal care products.

It recognizes that there is a significant volume of water associated with the sourcing of its raw materials. Its Sustainable Living Plan water metric includes water in the product and water used by consumers, but it does not currently include water used in agriculture since there is a lack of reliable data on agricultural water use for the wide range of crops that make up the ingredients in its products.

However, to increase its expertise in measuring water impacts, it is working with the Water Footprint Network (WFN) and the International Organization for Standardization (ISO) to extend the metric to include virtual water in its raw materials. Unilever cofounded the WFN in 2008 with organizations from the private sector, the International Finance Corporation, WBCSD, WWF, and UNESCO. It aims to develop a measurement framework that assesses the total water used across the life cycle of a product and the impact of that on water use. Unilever is now using WFN data from water footprint modeling to estimate the water footprint of its key agricultural crops. This has identified areas for further investigation.

The company calculated water footprints for two Unilever products (tea and margarine) as part of helping to develop the methodology and has been assessing the virtual water across Unilever's key agricultural raw materials. It is also contributing to LCA research to develop better data on virtual water in industrial products and has assessed water use across the life cycle of a laundry detergent, including the virtual water in ingredients.

Analyzing the water used to grow its agricultural raw materials is a complex task, but in 2011, it made good progress on estimating the water requirements of its key crops.

To see the data behind the Unilever Sustainable Living Plan, it has devised a Product Analyzer that shows the environmental impact of a selection of its products across their life cycle. This provides the greenhouse gases, water or waste impacts of a representative food, home or personal care product on a "per consumer use" basis. So, at the touch of a button, people can find out the greenhouse gas emissions associated with one cup of tea, or the water required for one wash with laundry powder, or the waste associated with one use of a roll-on deodorant.

Issues for Learning and Discussion

1. What are the standards Unilever used for assessing its water footprint?
2. What are the future challenges for Unilever related to water scarcity, and what should other organizations and governments do to face this issue?
3. What are the lessons Unilever learned from its water footprint product assessments?

BP: DEEP OIL SPILL CRISIS AND GREENWASHING

Related to This Case Study [8]: Chapters 14 and 20

Oil Spill Crisis Timeline   **December 1998** Construction begins on the Deepwater Horizon oil rig in Ulsan, South Korea, by Hyundai Heavy Industries Shipyard.

February 2001 The rig is delivered and valued at more than $560 million.

April 20, 2010 Explosion and fire on the BP-licensed Transocean drilling rig Deepwater Horizon in the Gulf of Mexico. Eleven people are reported missing and approximately 17 injured. A blowout preventer, intended to prevent release of crude oil, failed to activate.

April 22 Deepwater Horizon rig sinks in 5000 feet of water. Reports of a five-mile-long oil slick. Search-and-rescue operations by the US National Response Team begin.

April 23 The US coast guard suspends the search for missing workers, who are all presumed dead. The rig is found upside down about a quarter mile from the blowout preventer. A homeland security department risk analysis says the incident "poses a negligible risk to regional oil supply markets and will not cause significant national economic impacts." White House press secretary Robert Gibbs says: "I doubt this is the first accident that has happened and I doubt it will be the last."

April 24 Oil is found to be leaking from the well. A homeland security report on critical infrastructure says the problem has "no near-term impact to regional or national crude oil or natural gas supplies."

April 25 US coast guard remote underwater cameras report the well is leaking 1000 barrels of crude oil per day (bpd). It approves a plan for remote underwater vehicles to activate a blowout preventer and stop the leak.

April 26 BP's shares fall 2% amid fears that the cost of cleanup and legal claims will hit the London-based company hard. Roughly 15,000 gallons of dispersants and 21,000 feet of containment boom are placed at the spill site.

April 27 The US departments of interior and homeland security announce plans for a joint investigation of the explosion and fire. The coast guard announces it will set fire to the leaking crude to slow the spread of oil in the Gulf.

April 28 The coast guard says the flow of oil is 5000 barrels of crude oil per day, five times greater than first estimated, after a third leak is discovered. Controlled burns begin on the giant oil slick. BP's attempts to repair a hydraulic leak on the blowout preventer valve are unsuccessful.

April 29 President Obama talks about the spill at the White House, his first public comments on the issue. He pledges "every single available resource," including the US military, to contain the spreading spill and also says BP is responsible for the cleanup.

Louisiana declares a state of emergency due to the threat to the state's natural resources, as the oil slick approaches land.

April 30 An Obama aide says no drilling will be allowed in new areas until the cause of the Deepwater Horizon accident is established. The US justice department announces that a team of lawyers is monitoring the spill. Safety inspections of all 30 deepwater drilling rigs and 47 deepwater production platforms are ordered. BP chairman Tony Hayward says the company will take full responsibility for the spill, paying for all legitimate claims and the cost for the cleanup.

May 1 The coast guard announces the leak will affect the Gulf shore.

May 2 US officials close areas affected by the spill to fishing for an initial period of 10 days.

Obama visits the Gulf coast to see cleanup efforts first hand. BP starts to drill a relief well alongside the failed well. An additional 30 vessels and 1000 responders are deployed to the Gulf coast.

May 4 BP executives face Congress in a closed session, as the White House backs a Senate proposal to increase the limit on liability payouts from $75 million to $10 billion (£6.5 billion) for the cost of a spill.

May 5 BP successfully attaches a valve to the end of the broken drilling pipe at the Macondo well in a bid to end the flow of oil into the US Gulf. BP says one of the three leaks has been shut off by capping a valve, but that would not reduce the amount of oil gushing out. Officials conduct controlled burns to remove oil from the open water.

May 6 BP confirms the arrival of three huge containment domes designed to collect much of the 5000 barrels of crude oil per day leaking into the US Gulf from the Macondo blowout. The department of justice asks Transocean to preserve evidence in connection with the explosion and sinking of the rig.

Later On

BP admitted guilt on 14 criminal charges and agreed to pay a historic $4.5 billion (£2.8 billion) penalty in connection with the fatal explosion of its rig and the catastrophic oil spill.

The payments include $4 billion for criminal charges and $525 million to security regulators. BP will plead guilty to 11 felony counts of misconduct or neglect in relation to the deaths of 11 men aboard the Deepwater Horizon when the rig blew up and sank in April 2010, as well as misdemeanor counts under the clean water and migratory bird acts. The company will plead guilty to lying to Congress. "This marks the single largest criminal fine—$1.25 billion—and the single largest total criminal resolution—$4 billion—in the history of the United States," the attorney general, Eric Holder, told a news conference in New Orleans. "I hope this sends a clear message to those who would engage in this kind of reckless and wanton misconduct that there will be a significant penalty to be paid."

Three BP officials were also charged, in addition to the charges against the company. Donald Vidrine and Robert Kaluza, the two senior managers aboard the Deepwater Horizon, faced charges of manslaughter and of negligence in supervising the pressure tests on the well.

"In the face of glaring red flags indicating that the well was not secure, both men allegedly failed to take appropriate action to prevent the blowout," said assistant attorney general Lanny Breuer.

David Rainey, BP's former vice-president for exploration in the Gulf of Mexico, was charged with obstruction of Congress and lying about how much oil was gushing from the well.

The criminal settlement does not settle all of the claims against BP for the April 2010 blowout of the Deepwater Horizon and the subsequent oil spill.

BP is still on the hook for environmental damage to the Gulf of Mexico and could face up to $21 billion in penalties under the clean water act for restoration costs to waters, coastline, and marine life. Its chairman, Carl-Henric Svanberg, said BP believed the settlement was in the company's best interests. "We believe this resolution is in the best interest of BP and its shareholders," he said. "It removes two significant legal risks and allows us to vigorously defend the company against the remaining civil claims."

Green Rebranding

In late July 2000, BP launched a massive $200 million public relations and advertising campaign, introducing the company with a new slogan—"Beyond Petroleum"—and a green and yellow sun as its logo. The campaign was handled by Ogilvy & Mather Worldwide, one of the major advertising companies that also owns a slew of PR companies. Around the world the company took out full-page color advertisements in major magazines.

Ogilvy and BP later won the *PRWeek* 2001 "Campaign of the Year" award in the "product brand development." One of the advertisements run in the *International Herald Tribune* in November 2000 stated "Beyond...—means being a global leader in producing the cleanest burning fossil fuel. Natural Gas; means being the first company to introduce cleaner burning fuels to many of the world's most polluted cities; means being the largest producer of solar energy in the world; means starting a journey that will take a world's expectations of energy beyond what anyone can see today."

In a column for CorpWatch, researcher Kenny Bruno dissected the advertisement. "BP's re-branding as the 'Beyond Petroleum' company is perhaps the ultimate co-optation of environmentalists' language and message. Even apart from the twisting of language, BP's suggestion that producing more natural gas is somehow akin to global leadership is preposterous. Make that Beyond Preposterous," he wrote.

- While noting that BP was indeed the largest producer of solar energy, Bruno pointed out that was achieved by spending $45 million in 1999 to buy Solarex, which was dwarfed by the $26.5 billion it spent to buy ARCO to expand its oil portfolio.
- As for the claim that BP was starting a journey that would reshape public energy expectations, Bruno was scathing: "Pretentious stuff for a company serving mainly oil and gas, with just a sliver of solar on the side. Make that Beyond Pretentious."

The rebranding—undertaken in the wake of major controversies in Europe over Shell's role in Nigeria and its ill-fated attempt to dump the disused Brent Spar oil platform in the ocean—was aimed at differentiating BP from its rivals. Associate creative director with Ogilvy on the campaign, Michael Kaye, told the *New York Times* the campaign was aiming to communicate "BP can be a friend—listening to consumers, speaking in a human voice."

One of BP's PR advisers was Peter Sandman. While it is unknown whether he specifically advised BP on their rebranding project, nonetheless he described it at an Australian mining industry conference as an example of a company adopting the persona of being a "reformed sinner."

Sandman told his audience that this "works quite well if you can sell it. …'Reformed sinner,' by the way, is what John Brown of BP has successfully done for his organization. It is arguably what Shell has done with respect to Brent Spar. Those are two huge oil companies that have done a very good job of saying to themselves, 'Everyone thinks we are bad guys. …We can't just start out announcing we are good guys, so what we have to announce is we have finally realized we were bad guys and we are going to be better.' …It makes it much easier for critics and the public to buy into the image of the industry as good guys after you have spent awhile in purgatory."

With raised expectations about corporate behavior—and especially oil companies—BP's move not only sought to distance itself from its more notorious European counterparts but the brasher American oil companies—such as Exxon—which were fiercely opposing moves to curb greenhouse gas emissions.

Group vice-president for marketing for BP, Anna Catalano, told the *New York Times* that BP is "the company that goes beyond what you expect from an oil company—frank, open, honest and unapologetic."

BP also sought to cultivate "moderate" environmental groups in a series of "partnerships" with groups like the National Wildlife Federation. However, the trap for companies such as BP is that big-spending promotional campaigns often raise expectations that the organization is incapable of meeting. Where corporate PR is often adept at explaining away infringements or accidents to human error or failed equipment, considered corporate policy that is at odds with public expectations is harder to explain away.

BP's corporate rebranding was subject to skeptical review among activists and some mainstream media. In *Fortune* magazine, Cait Murphy cuttingly wrote of BP's billboards touting its involvement in renewable energy: "here's a novel advertising strategy—pitch your least important product and ignore your most important one…If the world's second-largest oil company is beyond petroleum, *Fortune* is beyond words," she wrote.

One of BP's regional presidents, Bob Malone, told Murphy "the oil business has a negative reputation…We are trying to say that there are different kinds of oil companies."

"As for being 'beyond petroleum'…Malone concedes that BP is decades away. Somehow that didn't make the billboard," she wrote.

Pressure groups accuse BP of splashing out more on advertising its environmental friendliness than on environmental actions.

- BP launches new greenwashing initiative while simultaneously calling for a big increase in CO_2 levels (500–550 parts per million):

"Based on current scientific opinion, we believe that it's realistic to promote actions designed to stabilize carbon dioxide concentrations at around 500–550 parts per million."

- In March 2007, the Australian reported that "Internal documents have revealed that BP successfully lobbied against tighter environmental controls by regulators in Texas, saving $US150 million in monitoring and equipment upgrades before the 2005 fatal refinery explosion."

Rebranding did not change many underlying BP activities. While BP's rebranding program may have reassured some of its critics, others remained unpersuaded. At its annual general meeting in April 2001, BP was challenged about its interests in projects spanning from Tibet, the Sudan and the Arctic National Wildlife Refuge.

- A resolution urging BP to disinvest from its shareholding in Beijing-controlled PetroChina—which plans an oil pipeline through Tibet—was opposed by the company, gaining support from only 5% of shares voted at the meeting. Speaking in support of the resolution, Stephen Kretzmann, from the International Campaign for Tibet, suggested that BP's slogan "Beyond Petroleum" should be changed to "Beijing's Partners" or "Backing Persecution." According to a report in *The Guardian*, he accused BP of "utilizing every arcane and legalistic tool to stifle debate on the matter." BP's chairman, Peter Sutherland, dismissed the concerns. "Disinvesting from PetroChina means, in reality, departing from China, which would be a mistake, and would be wrong," he told shareholders.

- Another resolution proposing that the company do more about climate change was also opposed by the board and defeated, gaining support from 7.5% of proxy voters. Sutherland told the meeting: "there have been calls for BP to phase out the sale of fossil fuels. We cannot accept this, and there's no point pretending we can."

- While as part of its rebranding program the company has touted its "ethics" policies, one shareholder activist attending the meeting challenged the directors to nominate a country that the company had decided to avoid because of human rights abuses. "After a long pause its chief executive, Sir John Browne, said it would be 'uncivil and inappropriate' to mention any no-go nations," *The Guardian* reported.

- BP's business ethics were also challenged when in June 2001 the London newspaper, *The Sunday Times*, revealed that both BP and Shell acknowledged that they hired a private intelligence company with close ties to the British spy agency, MI6, to collect information on campaigns by Greenpeace and the Body Shop.

The newspaper revealed that German-born Manfred Schlickenrieder was hired by Hakluyt, a private intelligence agency, to report on Greenpeace campaigns against oil developments in the North Atlantic. Schlickenrieder posed as a film maker working on films sympathetic to activist groups.

According to *The Sunday Times*, the former deputy chairman of BP, Sir Peter Cazalet, helped to establish Hakluyt, and former chairman of Shell, Sir Peter Holmes, is president of its foundation. In May 1997, the head of Hakluyt, Mike Reynolds, asked Schlickenrieder whether Greenpeace was planning to shield its financial assets from court orders in the event of it being sued by an oil company. Two months later, Greenpeace occupied the BP oil rig, the Stena Dee, in the Atlantic. BP sued Greenpeace for £1.4 million in damages and succeeded in gaining an injunction freezing the group's bank accounts while the occupation lasted. After police evicted Greenpeace campaigners from the rig, BP dropped its legal action and the freeze on the bank accounts was lifted.

"BP countered the campaign in an unusually fast and smart way," Greenpeace Germany spokesperson Stefan Krug told the German daily Die Tageszeitung. As Eveline Lubbers noted in PR Watch, "since BP knew what was coming in advance, it was never taken by surprise."

In other areas though, BP has made some concessions to public pressure. In early 2002, Lord Browne, who was then company chairman, announced that the company would no longer make donations to political parties anywhere in the world. In a speech to the Royal Institute of International Affairs, Browne said: "we have to remember that however large our turnover might be, we still have no democratic legitimacy anywhere in the world…We've decided, as a global policy, that from now on we will make no political contributions from corporate funds anywhere in the world."

Yet, in 2008, BP spent over half a million dollars on its US Political Action Committee. BP stated that it will continue to participate in industry lobbying campaigns and the funding of think tanks. "We will engage in the policy debate, stating our views and encouraging the development of ideas—but we won't fund any political activity or any political party," he said. In response to a question, Browne said that over the long term donations to political parties were not effective.

In August 2007, *Advertising Age* reported that BP had received "a permit from the state of Indiana to dump more toxic discharges from its Whiting, Ind., refinery into Lake Michigan." The permit, "which allows BP to dump 54% more ammonia and 35% more suspended solids" in the Great Lake, has "enraged" Chicago officials and "raised the specter of consumer boycotts." Chicago's chief environmental officer remarked, "We'd like to have [BP] live up to their advertising."

AdAge called BP's move "the cardinal sin of touting an environmentally conscious image in marketing—the central focus of BP's advertising for the past several years—and failing to live up to the message." A company spokesman said BP had "started advertising in regional newspapers…to clear up misconceptions about the issue."

BP later pledged it wouldn't increase its dumping into Lake Michigan. The pressure on the company was such that "Bob Malone, chairman of BP America, flew to Chicago to deliver the news personally to Mayor Richard Daley, one of several politicians who said the company's initial plans were unacceptable to the public," reported the *Chicago Tribune*.

In late 2007, BP also decided "to invest in the world's dirtiest oil production in Canada's tar sands," reported *The Guardian*. BP's investment in "the Alberta tar sands, which are said to be five times more energy-intensive to extract compared to traditional oil," prompted Greenpeace Canada to accuse the company of "the biggest environmental crime in history." BP's former chief executive, John Browne, "had said BP would not follow Shell into tar sands as he established an alternative energy division and pledged to take the group 'beyond petroleum.' The new boss, Tony Hayward, has pointed the corporate supertanker in a new direction although his public relations minders insist BP remains committed to exploring the potential of renewables," concluded *The Guardian*. Despite this pledge, BP Solar has recently cut over 600 of jobs from its manufacturing division, in what it claims is a "cost-saving move."

Issues for Learning and Discussion

1. What was the impact of the crisis on BP's brand reputation and share value?
2. Were the penalties paid worth the lack of investment in health, safety, and a more advanced sustainability strategy?
3. Was the BP rebranding campaign a greenwashing initiative and why?
4. Was the BP rebranding campaign initiative aligned with its sustainability progress?

REFERENCES

1. This Case Study was prepared by Heidi Owens, Ph.D., for The Natural Step Network. Available at http://www.naturalstep.org/en/search/node/case+studies+IKEA. Accessed 2013 Jul 4.
2. Reichert J. *IKEA and the Natural Step*. Charlottesville: Darden Graduate School of Business Administration, University of Virginia; 1996.
3. Moberg A. From Introduction of IKEA and the Environment, Feb 24, 1993.
4. This Case Study has been created by the Centre for Sustainability and Excellence (CSE) and it is based on the 2010 LBG CSR report and its work with LBG.
5. a. Apple asks outside group to inspect factories, *New York Times*, February 14, 2012, p. A1. b. Critics question record of monitor selected by apple, *New York Times*, February 14, 2012, p. B1.
6. Rim Sustainability Report 2012. Available at http://ca.blackberry.com/company/about-us/corporate-responsibility/product-sustainability.html. Accessed 2013 Jul 4.
7. Unilever Sustainability Websites. Available at http://unilever.com/sustainable-living/water/why/index.aspx and http://www.unilever.com/sustainable-living/water/footprint/. Accessed 2013 Jul 4.
8. Available at www.Sourcewatch.org and http://www.guardian.co.uk/environment/2012/nov/15/bp-deepwater-horizon-gulf-oil-spill. Accessed 2013 Jul 4.

Note: Page numbers in *italics* refer to Figures; those in **bold** to Tables.

Practical Sustainability Strategies: How to Gain a Competitive Advantage,
First Edition. Nikos Avlonas and George P. Nassos.
© 2014 John Wiley & Sons, Inc. Published 2014 by John Wiley & Sons, Inc.